HISTORIC ARMIES AND NAVIES

The Indian Army

HISTORIC ARMIES AND NAVIES

GENERAL EDITOR: *Christopher Duffy*

The Army of Frederick the Great *Christopher Duffy*

in preparation

The Army of the Second Empire *Richard Holmes*
D-Day Armies *E. Belfield*
The Jacobite Army of the '45 *Count Nikolai Tolstoy*
Maria Theresa's Army *Christopher Duffy*
Nelson's Navy *Michael Orr*
Parliament's Armies (1642–4) *Charles S. P. Kightly*

HISTORIC ARMIES AND NAVIES

The Indian Army

The Garrison of British Imperial India, 1822-1922

T. A. HEATHCOTE

DAVID & CHARLES
NEWTON ABBOT · LONDON
VANCOUVER

0 7153 6635 1

Set in Monotype Ehrhardt and New Clarendon Bold and printed in Great Britain by Ebenezer Baylis and Son Limited, The Trinity Press, Worcester and London, for David & Charles (Holdings) Limited, South Devon House, Newton Abbot, Devon.

Published in Canada by Douglas David & Charles Limited, 3645 McKechnie Drive, West Vancouver BC.

To my comrades of the
300th (Tower Hamlets) Light Air Defence Regiment
Royal Artillery (Territorial Army)
formed 1961
reduced 1967

Contents

Illustrations

Plates

All the photographs are by courtesy of the Records Department, National Army Museum, except for those on pages 17, 71, 90, 108, 125 and 126, which are from the Royal Military Academy, Sandhurst.

Maps

Tables

Prologue

Tribesmen in the hills, warlike and cruel, waiting to rob the merchants' caravans or harry the peaceful villagers. Rabble-rousers in the bazaar, stirring up the city mob to burn and loot. Ambitious princes in exquisite palaces, intriguing with the representatives of a certain northern power. If only they can unite, if only they can come by the modern arms and ammunition needed, if only they can wipe out the nearest British garrison, then this will be the signal for every malcontent in India to rise as one man, and drive the accursed English into the sea. In a week they will be in Delhi, in a month at the gates of Calcutta, and at the end there will not be a rupee or a virgin left between Cape Comorin and the Himalayas.

In their way stands the small but well-trained army of the Raj – sun-helmeted British soldiers and turbaned Indian sepoys. Its outposts are strung out in a long chain of tiny garrisons, some of which must fall in any sudden rebellion, but for those which hold out, there are relieving forces sent by Army Headquarters further back. Let the infantry come up, the guns get into position, the cavalry charge, and the rebel forces will dissolve into their component parts, each vainly seeking safety in flight.

It is in such a setting that the Indian Army is most familiar to the average reader. The purpose of this book is to fill in some of the details of that army, to show its soldiers, not as stereotypes, but as ordinary men, with hopes, problems and lives of their own; to show how they were organised, equipped, and commanded; how they lived, and how many of them died. It concentrates on the century when British power in India was at its height, beginning in 1822, when the British, already victorious in Europe after the defeat of Napoleon, established themselves as the paramount power in India by their conquest of the Marathas, and ending in 1922, after a victory in another great European war, when the Indian Army was reorganised to take account of a changed situation. Politically, the British were now committed to grant India self-government. Militarily, there were other changes. The soldier of 1914 would not have felt entirely strange had he suddenly been transported back in time to 1814. The cavalry still rode into battle with sword and lance. The artillery was still made up predominantly of light, horse-drawn, equipments. The infantryman still marched, with rifle and bayonet. But during World War I, with the advent of mechanisation, aircraft and automatic weapons, the pattern of warfare, and of armies, began to change rapidly. New problems brought new organisation, and the new situation brought to the Army additional types of officers and men, often different in ideas and background from those who had served before. This book then, is about the Indian Army in its classical period, the army which for a hundred years was the garrison of British Imperial India.

1

The Historical and Administrative Setting

The country

Of all the territories under British imperial rule, the Indian Empire made the deepest impression on the popular imagination. Unlike the other great dominions, which prior to the coming of the Europeans were inhabited only by savages or at best by backward tribal peoples, India had an ancient and sophisticated civilisation of her own. The very word, India, conjures up a series of brilliant images, kaleidoscopic in their variety – bejewelled rajas and maharajas in their opulent palaces; khans and amirs in the northern hills; ancient Hindu temples with strange gods; idols attended by priests in flowing robes; mosques and minarets, with the faithful of Islam called to prayer; teeming cities, older than Athens or Rome; villages, the homes of patient cultivators. The natural, as well as the man-made, sights of India seemed equally strange and exotic. The Himalayan mountains, highest in the world; the dry, boulder-strewn hillsides of the north-west; the jungles and hills of Central India; the great plains and mighty rivers; the palms and coral beaches of the southern peninsula; all flash before the imagination like scenes from a travelogue. British public opinion was curiously slow to wake to the nature and extent of British possessions in the 'East Indies', despite the material contribution which they made to British wealth, and despite the constant policy of successive governments to ensure, by diplomacy and armed force, that they were well secured. Yet, once this awareness came, India loomed large in British folk consciousness, as 'the brightest jewel in the British crown'. There developed a certain national, even racial, pride that a small sea-girt island, set in the grey Atlantic with a population of less than forty millions, could conquer an empire the size of Europe, and hold it with an army of 75,000 British soldiers and 150,000 Indian troops under British officers.

The Indian sub-continent covers an area of about 1,600,000 square miles, with every extreme of climate, terrain, and vegetation. Its population, over 400 millions, is made up from a wide variety of races, languages, and cultures. Its earliest known civilisation is that of the Dravidians or Tamil-speaking people, who were driven from the cities of the great Indo–Gangetic plains, about 1500 BC, by invading Āryans from the north-west. Some of the Dravidians remained as the serfs of, or collaborators with, the new conquerors. Others migrated southwards, dispossessing the primitive aboriginal tribes, and driving them into the hills and jungles. Sometimes they were themselves driven into the jungles by subsequent waves of Āryans, descendants of the first invaders.

The newcomers, who spoke an Indo-European language, Sanskrit, established themselves along the Ganges valley. Their religion was originally a simple pantheism, much like that of the early Greeks and Romans. It merged with the mystic cults of the original inhabitants to form a broad span of beliefs and philosophies which is given the generic name of Hinduism. 'Hindu' was in fact the term given by the later Muslim invaders to both the people and the religion of the country which began, for them, at the Indus, and was called Hind, or Hindustan. Hindu culture proved strong enough to spread southwards and be adopted by the Dravidians, who had no affinity at all with the early Āryans. It also proved vital enough to withstand successive waves of other invaders from the north-west in the ensuing centuries. Bactrian Greeks, Scyths, Persians, Turks and Mongols were all in turn absorbed into the Hindu system. By the fifth century AD the whole area between the Ganges and the Indus, a vast arid and hilly tract called Rajasthan, was held by descendants of Central Asian invaders, all of whom had identified themselves with Hindu culture.

More invaders from Central Asia and Afghanistan, zealous converts to the faith of Islam, raided the Hindu kingdoms of northern India in the early eleventh century AD. By 1206 all the northern plains were ruled by the Muslim sultans of Delhi. These in turn were supplanted by new invaders, called Mongols, or *Mughāl*, and by the sixteenth century AD the Muslim Mughal emperors ruled nearly the whole of India.

The Muslims, unlike earlier invaders, were fortified against the all-pervasive embrace of Hinduism by their fiercely monotheistic faith. No dialogue was possible between true believers and the idolatrous Hindus, who committed the cardinal sins of giving God a partner (polytheism), of denying His existence (unbelief), or of believing the wrong thing about Him (heresy). Islam was identified as the religion of the conquerors, and although, in the states still ruled by Hindus, Muslims were at a disadvantage, in most of India even the humblest Muslim could expect his ruler to ensure that he enjoyed advantages over the unbelievers.

The British period

The British Empire in India grew from a collection of mere trading stations maintained by the 'Governor and Company of Merchants of London trading to the East Indies'. This body, granted its charter by Queen Elizabeth I on the last day of January 1600, enjoyed a monopoly of English trade in eastern seas. Like the chartered companies of other European nations, the East India Company had the power, economic and legal, to defend its interests against unfriendly foreign competitors or local Indian rulers. As the Mughal Empire declined during the late seventeenth and early eighteenth centuries, the European companies became involved in the local quarrels of its successor states. In return for the support of a company's money, ships, and soldiers, a local warlord, if successful in becoming an established ruler, would grant it (and its servants) valuable commercial privileges.

By 1764 British victories in Bengal and the Carnatic, over the French Compagnie des Indes and its Indian allies, had brought tracts of territory in south and east India under the control of British nominees. Much of this was subsequently brought under direct British rule.

The East India Company thus ceased to be a mere trading concern, and became itself an Indian power. It entered the political life of the sub-continent, conducted diplomacy, established alliances, and took part in wars. One by one the major Indian states came under British control, either conquered and incorporated into British India, or allowed to remain as satellites, bound to the British as subordinate allies. By 1822, after the defeat of the powerful Maratha Confederacy which had controlled most of western and central India, the British were established as the paramount power in all India.

The outlying states were dealt with in a series of frontier wars. Sind, at the lower end of the Indus Valley, was conquered in 1843. The fertile and populous plains of the Punjab were conquered in the two Sikh Wars of 1845 and 1848. Three wars against Burma between 1824 and 1886 led to the annexation of that country. Within the frontiers, several large states were bloodlessly annexed by the British in the 1850s, using the Mughal doctrine that subordinate states lapsed to the paramount power if their ruler failed to produce natural heirs. Only on the hills of the barren North-West Frontier did the wave of British conquest break. Two invasions of Afghanistan, in 1839 and 1878, both had the same result; initial victories, followed by popular uprisings, and humiliating withdrawals.

THE MACHINERY OF GOVERNMENT

The growth of the Company's power in India led to a gradual extension of parliamentary control over its affairs. It was thought wrong for Englishmen to have power over armies and navies and great provinces, without being checked by the English Parliament. Public opinion became alarmed by stories of the rapacity of servants of the Company in India, and by the conspicuous wealth and influence which they exhibited on their return to the United Kingdom. From 1774 the Company's dividends were fixed by Act of Parliament, not its own Directors. With each renewal of the Charter Acts which authorised the Company to hold power in the East, the Company's authority was reduced.

The 1784 Act brought all the Company's military and political activities directly under a newly formed Board of Control. This was virtually a ministry of state, with a cabinet minister, the President of the Board, at its head.

The Company remained as the agent through which British possessions in India were ruled. The most important reason for this was the question of patronage, the right to appoint government officials. Public offices, or 'places', were regarded as valuable property. In contemporary politics, a government in the United Kingdom was expected to reward the past (or ensure the future) loyalty of Members of Parliament who supported it in the House of Commons, by appointing their nominees to suitable jobs in government employment. The British civil service was kept very small by parliamentary scrutiny of the official estimates, since the salaries of civil servants are paid for by the proceeds of taxation, and the supplies have to be voted for each year. This kept down the amount of 'patronage' which the king's government had at its disposal, and so limited the amount of support it could obtain by this means. But the number of appointments in the Company's Indian civil service was increasing, and so was the number of officers in its army (who were nominees of the Company and not, as in the British Army, men who had purchased their commissions from their predecessors). It was clear that any government with the patronage of India at its disposal would be able to gain enough support in Parliament to keep itself in office indefinitely, and it was in order to avoid such a situation that the Company was kept in existence.

The military and civil appointments of British India were thus in the gift of the twenty-four men who made up the Court of Directors of the East India Company. Theoretically elected by the Proprietors of the East India Company's stock, the Court was in fact a co-optive body. It used its powers wisely, however, and ensured that each branch of the Company's activities, commercial, revenue, military, or maritime, was represented among its members. Thus there were always at least two directors who had served with the

Company's army in India. On average, each director could nominate every year three assistant surgeons, three civil servants, and twelve military cadets.

As well as acting as a sort of appointments board, the Court of Directors provided for the government of the day a valuable source of independent and up-to-date opinion on Indian affairs. All 'dispatches' to, and 'letters' from India passed through the Company's home headquarters, and the Board of Control took note of the Company's views in forming any policy decisions. The Court of Directors was divided into several committees, each served by permanent officials who advised the directors set over them. The Board of Control divided its business among several departments in the same way, each with its own staff of civil servants to assist the minister in his work.

The Indian Mutiny of 1857, in part a military rebellion by the Bengal section of the Company's native army, in part a popular revolt against the rapid extension of British rule, served as the occasion for the abolition of the Company's government. As early as 1833 it had ceased to be a trading concern. The main reason for its continued existence, the patronage question, had disappeared in 1853, when the directors surrendered their right to fill civil offices in India in favour of the principle of selection by open competitive examination. In the aftermath of the Mutiny, the British Government decided that from then on India would be governed directly in the name of the British Crown. The Company's servants, powers, and territories were transferred to the Crown in India. The President of the Board of Control became the Secretary of State for India. The home establishments of the Company and the Board were merged to form the India Office, and a Council of India, composed of men who had held high office in India, was set up to provide the minister with that Indian expertise formerly found in the Court of Directors.

In India, the last surviving ruler of the Mughal dynasty, the titular King of Delhi, was deposed by the British for his complicity in the Mutiny. Indian public opinion was somewhat shocked at this, since in strict constitutional terms the Company held its dominions in India by virtue of imperial Mughal *sanads* or charters, and by withholding tribute from the King of Delhi, and by ignoring his proclamations during the Mutiny, the British had been technically in rebellion against him. The removal of this ancient monarch left no rival to the British Crown in India, and in 1877 the extension of British sovereignty was completed by the Royal Titles Act, whereby Parliament authorised Queen Victoria to assume the additional title 'Empress of India'. But the British monarchy could not use its new-found title outside India. The adjective 'imperial', with its connotation of the brutal tyrannies of Habsburg, Hohenzollen, Bonaparte, or Romanov, was offensive to many Englishmen, and suggested a threat to their ancient hard-won liberties. Officers of the British Army, even in India, continued by custom to drink the health of 'the Queen'. Only officers of the local Indian service drank to the toast 'the Queen-Empress'.

The Government of India, although often depicted as a powerful autocracy, remained firmly accountable, in the final analysis, to the elected British Parliament. The Governor-General took his instruction from the Secretary of State for India, who, like all other cabinet ministers, had to answer to the House of Commons. Moreover, his acts were additionally circumscribed by the powers of the Council of India, which could veto any measures involving the expenditure of Indian revenues. In particular, the Indian Army could not be used outside Indian territory without express parliamentary sanction, thus retaining the constitutional check on the military resources of the Crown.

The India Office was divided into departments, each serving one of the committees

The advance of the 31st Foot at the battle of Firozshah, 22 December 1845, First Sikh War. Lithograph by J. Harris after Henry Martens

(above) *Sepoys with pack-mule, in service dress. Punjab infantry, 1904. Watercolour by A. C. Lovett;* (below) *Sepoys in full and service dress. Pioneer regiments, 1904. Watercolour by A. C. Lovett*

into which the Council of India was formed. Each department had its own permanent secretariat, which dealt with the chairman of its committee, who in turn had direct access to the Secretary of State in policy-making decisions affecting his department. The military department was staffed by serving officers of the Indian forces, and its permanent secretary was usually a major-general. There was always at least one retired general of the Indian service on the Council of India.

At the beginning of the eighteenth century, British possessions in India were divided into the three 'presidencies' of Bengal, Madras, and Bombay, each independent of the others, each with its own military and civil officials, and each under a 'Governor and President-in-Council' who was directly responsible to the Court of Directors in London.

In 1774 the Governor of Bengal was made Governor-General of Bengal, and the governors of the other two presidencies were made subordinate to him in matters of foreign policy. The control of the Governor-General was gradually extended, and by the 1853 Charter Act he was for the first time styled Governor-General of India, and his control over the entire field of British political, civil, and military policy in India was defined in clear and precise terms. After the transfer of the Company's dominions to the Crown in 1858, the Governor-General was commonly referred to as the Viceroy of India. Despite the magnificence of this latter title, and its frequent use by romantics or those with a taste for high-sounding expressions, the term was constitutionally incorrect. There was never any statutory authority for its adoption, and it never appeared in any of the warrants by which Governor-Generals held their office. The title was one of ceremony, used especially when the Governor-General acted as representative of the Crown. It was particularly suitable for use on occasions such as the Imperial Assemblage at Delhi of 1877, when the assumption by Queen Victoria of the title 'Empress of India' was declared to the assembled princes and chiefs of India. However, it was as Governor-General that he held the position of chief executive. He was accorded the appellation 'His Excellency' and a salary of two-and-a-half lakhs of rupees per annum. Governor-Generals were for the most part British politicians, sometimes men at the end of their careers, but sometimes men who could look forward to higher posts at cabinet minister level. They were appointed by the Crown, usually for a term of five years. Few had previous experience of India.

The Governor-General was not by any means the all-powerful figure that is depicted in many popular books about India. His policy was always subject to control by the British Cabinet, and he was always bound to carry out its instructions. Only the difficulty of communications allowed him a certain latitude in applying them. In the early part of the nineteenth century, as much as two years could elapse between the sending of a letter from the seat of government at Calcutta to London and the arrival of a reply. By the middle of the century, the Cape route had been replaced by a shorter one, with the mails being carried overland through Suez to the Mediterranean, and by rail from Marseilles to Calais. The introduction of reliable steamers and new railways, and the construction of the Suez Canal in the 1860s, cut the time needed to get a reply from London to six weeks (plus the time spent in deliberation at the India Office), and the construction of the Indo–Iranian telegraph line, at the same time, brought India and England to within a few hours of each other. Nevertheless, the way in which policy was carried out was left to the Governor-General, with the ever-present sanction that a man who acted contrary to the spirit of his instructions could be recalled or forced to resign. Any attempt by a Governor-General to act as an independent ruler was firmly discouraged.

The Governor-General was advised by a Council. Members of Council were appointed, like the Governor-General himself, by the Crown, for a period of five years. In exceptional circumstances the Governor-General could overrule his Council, but this power was rarely exercised. In normal circumstances questions were decided by a simple majority of the Council.

The Indian Council Act of 1861 sanctioned the adoption by the Governor-General's Council of a 'portfolio' system. Each member of the Council, which had until then been intended to deliberate on all subjects of official policy, was given responsibility for his own department. Thus there was a member for the Home Department (usually a former Indian civil servant), who was in effect Minister of the Interior, a Law Member (an experienced barrister or advocate), a Finance Member (usually a banker or economist) and so on. There was no separate member for the Foreign Department, which was a special responsibility of the Governor-General in addition to his other duties. The Member for the Military Department was an officer of the Indian service, usually a major-general. Unlike some of the other members, he was a man who could expect to go on to hold other high appointments in his profession. Like them, he was appointed by royal warrant, and held office independently of any other authority. He was not permitted to command troops while a Member of Council.

The Government of India had divided its business into several departments long before the portfolio system appointed Members of Council to preside over them. Each department had its own permanent secretariat. Dispatches from London were received in India and read in the appropriate department, where its secretary and his staff would collect opinions, assemble papers, and pass them on for a decision by the Governor-General in Council. Constitutionally, all secretaries of departments were secretaries to the Government of India as a whole, and possessed the right of access to the Governor-General on any official subject. The Secretary of the Military Department was also a serving officer of the Indian forces, usually with the rank of major-general. Thus there were two senior officers, as well as the Commander-in-Chief, India, who had the right to discuss military subjects with the Governor-General.

The Military Department dealt with the administration and supporting services of the Army. Its task was to supply the Commander-in-Chief, India, with the men, money, and munitions needed by the Army to enable it to carry out the Government's policies. It controlled the ordnance depots and the factories producing weapons and equipment. It supervised the supply and transport services, the medical and veterinary departments, and the garrison works engineers. Rates of pay, establishments, stores, and buildings were all subjects controlled by the Military Department rather than the Commander-in-Chief. The finances of India were always in difficulties, and the Military Department was especially responsible for ensuring that the estimates for expenditure were not exceeded.

The annual movement of British units, on relief, between the United Kingdom and India was a typical subject that had to be judged nicely for financial reasons. If the British War Office did not take back the troops at the right time, the Indian exchequer would be bankrupted. If India did not accept troops at the right time, they would be surplus to the British establishment for which Parliament had voted supplies, and would have to be disbanded.

The Commander-in-Chief, India, was also on the Governor-General's Council, as an

Extraordinary Member. Normally he took part only in discussions relating to military matters, or to foreign policy involving the preparation for likely hostilities.

This division of responsibility between the Commander-in-Chief, India, and 'The Government of India in the Military Department' led to a major controversy in 1904–5. The arrangement had worked well enough until then, and probably would have continued to do so but for the personalities of the two men involved. Both were ambitious, proud, unscrupulous, and stubborn. One was Lord Curzon of Kedleston, Governor-General of India, and the other was Lord Kitchener of Khartoum, Commander-in-Chief, India.

Curzon secured the appointment of Kitchener to India with a view to bolstering his own prestige. He savoured the idea of having at his side a general whose triumphs in the Sudan and South Africa had made him one of the most famous soldiers of the British Army. He found, however, that Kitchener had his own plans, and his own prestige to use in carrying them into execution. Under Curzon, the Military Member, Major-General Sir Edmund Elles, had visited the North-West Frontier, and planned a number of alterations in the disposition there, in accordance with Curzon's own ideas. Kitchener was not prepared to see any rival to his own authority, and was determined to dispose of Sir Edmund Elles and his whole department.

He therefore proposed that the Government of India should have only one adviser on military subjects of any description: himself. The Commander-in-Chief, India, he said, should be in charge not only of operations, training, and discipline, but also of the supply of its material and administrative needs. It was contrary to all notions of military discipline that an officer, junior in rank to the Commander-in-Chief, could advise against his proposals, and freely discuss in Council the merits or otherwise of his plans. Curzon argued that to concentrate all these functions into the hands of one man would create a military autocracy and deny the Government of India an independent source of military advice.

A bitter feud occurred between the two men and their partisans. Both intrigued with their friends in the Court and Cabinet in England. Kitchener corresponded on the subject with the British War Office. Curzon pointed out that all official business must go through the Government of India. Kitchener thereafter sent his telegrams in code. Curzon used his powers as Governor-General to call secretly for the Army code book, and personally deciphered the messages before allowing them to be dispatched. Kitchener was supported by most of the officers of the Indian Army. Curzon had alienated the British Army by punishing regiments which tried to hush up assaults by their soldiers on Indians. He offended all officers by making it clear that he had a low opinion of military men in general, and those of the Indian service in particular. 'They seem to me congenitally stupid,' he wrote to the Secretary of State for India. 'Their writing is atrocious . . . the majority fill me with despair.'

Eventually a compromise seemed to have been reached. An Army Department was set up, headed by the Commander-in-Chief as a full Member of Council, to deal with all military business of the Government of India except stores and supplies. The Military Department was renamed the Military Supply Department, and its powers were curtailed. However, the Military Supply Member was still to be a serving general, and Kitchener insisted on the right to nominate him. This negated the principle for which Curzon had been fighting, the right to have independent advice. He threatened to resign, and to his chagrin his resignation was accepted. In 1909 Kitchener secured the abolition of the

Military Supply Department, and sat as the only soldier in the Government of India, as Commander-in-Chief and Member for the Army Department.

This system was in force when the Indian Army was engaged in the campaigns of World War I. It proved to be completely unsuitable for the complexities of major war. Kitchener's quest for personal aggrandisement had resulted in an administration that was unwieldly, over-centralised, and impossible for one officer to direct. The wretched General Sir Beauchamp Duff (C-in-C India 1914–16), an incompetent even when compared with his contemporaries, so far from acting as the commander-in-chief of a great army in fact lived the life of a hermit clerk. The collapse of the Indian Army's logistical arrangements in the Mesopotamia expedition resulted in a major scandal, in which Kitchener's system was completely discredited.

The Governor-General also had a personal military staff officer, the Military Secretary. This officer had the special task of dealing with the selection, promotion and retirement of officers of the Indian Army serving with local forces, under the control of the Government of India in the Foreign Department. In this his duties corresponded with those of the Military Secretary to the Commander-in-Chief, India, who dealt with the appointment of officers in the regular forces. Both military secretaries were usually experienced officers, with the rank of lieutenant-colonel or colonel. Some Governor-Generals regarded their military secretaries almost as major-domos, and entrusted them with little more than such matters as the superintendence of the young officers employed as aides-de-camp, and the organisation of garden parties. Lord Curzon even expected his military secretary to stand behind the vice-regal chair at meal times. In other cases, however, the post was one of greater influence. After the abolition of the separate Military Member, much of the correspondence between the Governor-General and the Commander-in-Chief, India, was carried on through their respective military secretaries.

One of the most influential of such officers was Colonel George Colley, Military Secretary to the Governor-General between 1877 and 1879. He was an able and intelligent officer, who had studied his profession carefully, and was one of the 'Wolseley ring'. Sir Garnet Wolseley, the 'very model of a modern major-general', and the officers in his circle, were noted especially for their scientific approach to the problems of command and staff duties, and were associated with the improved prestige of the British Staff College, Camberley, from the 1860s onwards. Colley was selected by the Governor-General, Lord Lytton, with the advice of Wolseley and the Prime Minister, Disraeli, as being specially qualified to help in what Lytton was later to describe as a struggle against 'the powers of military darkness' (i.e. the C-in-C India and the Military Member).

Lytton was sent to India by Disraeli with the express task of improving the British position on the North-West Frontier. In following a more aggressive policy than had been previously adopted, he became involved first in tension, and finally in a war, with Afghanistan. Colley had complete faith in the value of the recently adopted breech-loading rifle as a weapon capable of dominating the battlefield. He argued that a single regiment of British soldiers, armed with breech-loaders and well supplied with ammunition, could march at will through the length and breadth of Afghanistan. This belief enabled him to supply the Governor-General with proposals that were attractively economical.

Colley had left India by the time the Afghans had disproved his theories. He died on Majuba Hill, as Governor and Commander-in-Chief of Natal, where his army suffered a disastrous defeat at the hands of the Boer settlers.

Many Governor-Generals could rely on their own military experience to supplement the views of their official advisers. Of the twenty-two Governor-Generals to hold office between 1819 and 1925, only nine had no personal experience of soldiering. Two of them held the posts of Governor-General and Commander-in-Chief, India, simultaneously, drawing both salaries. One was the Marquis of Hastings who had been a general in the American War of Independence; the other was Lord William Bentinck, who had held several command and staff appointments during the wars against Napoleon. Another veteran of the Napoleonic Wars was Sir Henry Hardinge. He served during the First Sikh War as second-in-command of the army in the field. In fact his services were of doubtful value, as on occasions he countermanded, on political grounds, orders which had been drawn up by his commander-in-chief. The Marquis of Dufferin, although a civilian, had served as a volunteer with the naval expedition against Bomarsund during the Crimean War, and had fought on foot in the trenches with the French attacking force. He had also been Under-Secretary of State at the War Office from 1866 to 1868, and chaired the 1871 Commission of Enquiry into Military Education. Lord Northbrook, Governor-General from 1872 to 1876, was another politician who had been Under-Secretary of State for War, and was indeed Dufferin's successor in that office. Lytton's successor, the Marquis of Ripon, had been honorary Colonel of the West Yorkshire Volunteer Rifles. Lord Chelmsford was in the reserve Territorial Force. He went to India in 1914 as a captain with the Dorset Regiment, and was appointed Governor-General a year later. The Earl of Minto held a commission in the Foot Guards between 1867 and 1870, served as a volunteer with the French Army during the Commune, and attached himself to Lieutenant-General Sir Frederick Roberts' staff during the Second Afghan War. Even John Lawrence, the civilian *par excellence*, the only Indian civil servant to become Governor-General of India, eventually donned uniform, to become honorary colonel of the 41st Middlesex Rifle Volunteer Corps, recruited from the workmen of the Royal Ordnance Factory at Enfield Lock, home of the Enfield rifle that was said to be a cause of the Indian Mutiny.

The governments of the two minor presidencies, Madras and Bombay, were organised in a similar way to that of the Government of India but on a smaller scale. Their governors were appointed by the Crown for a five-year term, and were, like the Governor-General himself, selected from the ranks of British politicians or of Indian civil administrators. They both had the right to correspond directly with the home authorities on matters of concern to their own presidencies. Each was paid a salary of one-and-a-quarter lakhs of rupees per annum. They were assisted by a council of two members, also appointed by the Crown, from the ranks of the Indian civil service. The business of government was divided into departments with permanent staff, but each Member of Council and Secretary presided over the affairs of two or more departments. Prior to 1895, each presidency had its own army, with its own commander-in-chief, who was responsible to the governor. Many governors had been regular or volunteer officers and some held high military rank.

2

The Organisation of the Army and the High Command

The term 'Indian Army' is used in this book to denote the whole of the land forces at the disposal of the Government of India. These forces were divided until 1895 into three separate armies, to each of which was attached a proportion of troops belonging to the British Army. These three armies, with their British Army components, were localised under the control of each of the three presidencies into which British India was divided, and were known as the Bengal, Madras, and Bombay armies respectively. The term 'Indian Army' was sometimes used in a more restricted sense to mean only the native troops and their officers, and other units employed directly by the Government of India. The whole force was sometimes referred to as the Army of India, or the Army in India, and this latter title was officially adopted in 1903 to mean all the military forces in India. The term 'Indian Army' was thereafter officially used in the narrow sense to refer to the troops recruited and based in India, and their European officers. To avoid the anachronism of using it in the restricted sense throughout the period covered by this book, it has generally been given the wider meaning, and terms such as 'Indian service' or 'Indian troops', to indicate the local forces, have been used instead.

The term 'British Army' was also used in different senses from time to time. Sometimes it referred to all the troops in the service of the British authorities in India, as opposed to troops of other, local, powers. Generally it referred to the United Kingdom's own land forces, of which at any given time a number were serving a tour of duty in India.

The Commander-in-Chief, India

The chief executive officer of the Indian Army was the Commander-in-Chief in the East Indies, usually known as the Commander-in-Chief, India. The holder of this appointment was in direct command of the Bengal Army, and exercised general control of the armies of Madras and Bombay. He was selected, at first, by the Directors of the East India Company on the advice of the Crown, and later by the Secretary of State for India. He was not subordinate to the Commander-in-Chief of the British Army, nor was he responsible to the British War Office, and all his official correspondence with these was through the Government of India. Although supreme authority over all the troops in India was vested in the Governor-General in Council, the Commander-in-Chief had a separate responsibility for the manner in which military operations were conducted, and for the efficiency and discipline of the troops which took part in them. The decision to undertake a particular campaign was one taken, on political grounds, by the Governor-General, but the Commander-in-Chief alone had the duty to decide what form the campaign should take, what forces were needed, what plans should be made, and what orders issued. Just as the Governor-General was the Crown's civil and political representative, the Commander-in-

Chief was its military representative. Appointment was usually for a term of five years, and carried the salary of one lakh of rupees per annum, with the appellation of 'His Excellency'. Commanders-in-Chief, India, were usually generals or lieutenant-generals in rank, and the vast majority were officers of the British Army. Between 1822 and 1922 only seven out of twenty-six came from the local Indian service.

It was the usual practice on the outbreak of major hostilities in India for the Commander-in-Chief to take the field at the head of any army. Thus, in the nineteenth-century, the Marquis of Hastings, who was Governor-General as well as Commander-in-Chief, personally commanded the Grand Army of Bengal, which defeated the Marathas in 1817; General Lord Combermere led the army of 21,000 men which captured the city and fortress of Bharatpur in 1825, and General Sir Hugh Gough commanded the Army of the Sutlej, in the First Sikh War in 1845–6, and the Army of the Punjab in the Second Sikh War of 1848–9. The last Commander-in-Chicf, India, to command a field army in person was Sir Colin Campbell (Lord Clyde) in the campaigns of the Indian Mutiny, 1857–8.

In campaigns beyond the frontiers of British India, the Commander-in-Chief, India, remained at his headquarters, and left the day-to-day conduct of operations in the hands of the generals commanding columns in the field.

This pattern of command differed from the British Army's system, in which the Commander-in-Chief, or (later) the Chief of the Imperial General Staff, was responsible only for putting an army into the field, and its subsequent discipline and supply. A British general commanding an overseas expedition, such as the Duke of Wellington in Spain, or Lord Raglan in the Crimea, dealt directly with a cabinet minister in the British Government, the Secretary of State for War and the Colonies, without reference to the Commander-in-Chief. Generals commanding Indian army expeditions dealt with the Government of India only through the Commander-in-Chief, India. But in both systems the proceedings of generals in the field were controlled by the political authorities at the head of the civilian governments.

Army headquarters

The Commander-in-Chief, India, had an establishment of senior staff officers to assist him. Until 1895, their duties, like his, were primarily connected with the command of the Bengal Army, and secondarily to correspond with their opposite numbers on the headquarters of the armies of Madras and Bombay. The two branches into which the staff were divided were those of the Adjutant-General, who was responsible for matters connected with the soldier as an individual, and the Quartermaster-General, who dealt with the soldiers' material needs. These posts were usually held by officers with the rank of major-general, who had under them a small number of Deputy Adjutant (or Quartermaster) Generals, Assistant Adjutant (or QM) Generals, and Deputy Assistant Adjutant (or QM) Generals. Deputies were usually colonels, Assistants were usually lieutenant-colonels, and Deputy Assistants, majors. The Commander-in-Chief's Military Secretary, usually a colonel, was the head of his personal staff, and had special responsibility for dealing with officers' appointments, promotions, exchanges and retirements. Senior commands and staff appointments required the approval of the Governor-General and the home authorities, and the concurrence of the British Commander-in-Chief or Army Council was required if an officer of the British Army was involved.

In 1903 a general staff was created, with the tasks of dealing with matters of overall military policy, the supervision of training in peace, the conduct of operations in war, and the distribution of the Army for internal security or external use. It was also responsible for drawing up plans for future operations and the collection of intelligence. Previously this had been a function of the Quartermaster-General's department, but until about 1890 no formal intelligence branch existed. Much reliance for information, about the country or enemy which the army was likely to meet, was placed on police reports, or spies, agents, and informers in the pay of the 'Political Service', which was the official title of the Indian diplomatic service.

The post of Chief of the General Staff (India) was held by a lieutenant-general. In 1917 the Adjutant-General's and Quartermaster-General's posts were upgraded to the same level, and at the same time the post of Military Secretary was regraded, to be held by a major-general.

The Commanders-in-Chief of Madras and of Bombay maintained similar establishments but on a smaller scale. Their staff officers were fewer in number, and usually one grade lower in rank and level of appointment. The commanders-in-chief themselves were usually officers with the rank of lieutenant-general. Under the general superintendence of the Commander-in-Chief, India, they were directly responsible to the governors of their own presidencies for the efficiency, discipline, and employment of their armies. The Commander-in-Chief, India, did correspond with them directly, but in constitutional terms they were subordinate to their own governors, who were subordinate to the Governor-General, who in matters of military policy was advised by the Commander-in-Chief, India.

The majority came from the British rather than the local Indian service. After the assumption of direct rule by the Crown in 1858, the custom generally (though not invariably) followed was that at least two out of the three chief commands were held by British Army officers, and that when an officer of the Indian service held a command, his successor was drawn from the British service.

Troops under the Foreign Department

As well as the three main armies, smaller forces were raised which were not under the immediate control of the Commander-in-Chief. These were the troops maintained by the Government of India in the Foreign Department. Such forces might be lent to the Commander-in-Chief in an emergency, but their primary function was the defence of the particular areas where they were permanently stationed.

Such forces included the subsidiary contingents paid for from the revenues of Indian princes who were subordinate allies of the British. The local ruler, in return for British protection, accepted the cost of maintaining the troops required to garrison and protect his territories. The troops were commanded by British officers of the Indian service, and were recruited from the same 'martial classes' as the Indian Army's soldiers. Thus the contingents were in no way subjects of the rulers whose states they were defending. To pay for the expense of equipping and training these troops to the same standard as the rest of the Army, the local ruler usually granted to the British the right to collect the revenue of nominated districts up to the required amount, and the British, in assuming their revenue administration, usually assumed the civil and judicial administration as well.

The frontier forces

Other troops not always under the Commander-in-Chief, India, were the two frontier forces, raised for the defence of Sind and the Punjab respectively. The Sind Frontier Force was the first frontier force to be formed as such, and guarded the southern part of the North-West Frontier of India. It was raised in 1846, and at first consisted of only one regiment, the Scinde Irregular Horse. A second regiment of Irregular Horse was raised in the same year, followed by a third in 1858. Two infantry battalions (Jacob's Rifles) were raised in 1858, and provided soldiers to man the Jacobabad mountain train, which was transferred to the artillery in 1876. They were based at Jacobabad (Khanghar), a settlement in the desolate plains of the frontier district of Upper Sind, where there was a constant danger of raids by tribesmen from the dominions of the Khan of Kalat across the border in Baluchistan. This force was part of the Bombay Army, and so was controlled by the Bombay Commander-in-Chief.

On the annexation of the Punjab in 1848 the British were faced with two major problems. The first was the existence of a large number of unemployed ex-soldiers, who had been demobilised, or thrown out of work, by the disbandment of the regular army of the state of Lahore after the British conquest. The second problem was the defence of the Punjab and its inhabitants from the raids of Pathan tribesmen, living in the independent and mountainous areas across the border. These tribesmen, like the Baluchis of the Upper Sind Frontier, derived a substantial part of their income, in times of disorder or weak government, from descending on the plains-dwellers, or passing caravans, and stealing their property, using violence against any who tried to resist them. The North-West Frontier of British India was a most unpleasant area, with a trying climate and a semi-desert mountainous terrain. Regular troops tended to regard it virtually as a penal colony, and service there was unpopular. The regular troops proved too slow to deal with the hit-and-run tactics of the Pathan robbers, and so the expedient was devised, following the Sind model, of forming a locally based frontier force. Some of the disbanded troops of the Sikh Army had already been formed into regular regiments of the Bengal Army. In 1849 five regiments of Punjab cavalry and five battalions of Punjab infantry were raised, and these combined with the four battalions of Sikh Infantry, raised in 1846–7 for frontier protection, to form the Punjab Irregular Frontier Force. This force also included the Corps of Guides (in which a cavalry regiment and infantry battalion were combined in one unit) and an artillery component made up of two mountain trains, three light field batteries (these were reduced to form two mountain batteries in 1876) and a garrison company to man fixed defensive armament. The Scinde Camel Corps was transferred from the Bombay establishment in 1849 to become the 6th Punjab Infantry, and continued to carry the designation of Scinde Rifles, although it had never been part of the Sind Frontier Force.

This force came under the control of the Lieutenant-Governor of the Punjab, who was responsible to the Government of India in the Foreign Department for the conduct of its operations.

The Punjab Force was less successful in pacifying its frontier than the Sind Force had been, and fighting continued on the north-west border of the Punjab until the last days of British rule. In defending it against critics, John Lawrence, the first Lieutenant-Governor of the Punjab, argued that even the division's-worth of troops at his disposal was small in proportion to the area to be defended. Posts were therefore spaced at long intervals and their garrisons were weak. Large areas of the frontier were not garrisoned at all, and the posts generally were well back from the frontier. Settlements were close to the foothills, and flocks commonly grazed both sides of the borderline. It was impossible to disarm the tribesmen on the British side. Rather they were encouraged to go well-armed against robbers. The Sind Force, he admitted, was efficient, but it had to cover only one part of Upper Sind Frontier, and could be concentrated there, as the tribesmen elsewhere were peaceful and it was possible to disarm the local population.

Brigadier-General John Jacob, the first commander of the Sind Frontier Force, and Political Superintendent of the Upper Sind Frontier, felt that his greater success was attributable to moral rather than physical factors. In a memorandum on his proceedings in Upper Sind up to 1854, he summed up his approach to the subject.

> The highest moral ground always taken in all dealings with the predatory tribes, treating them always as of an inferior nature so long as they persist in their misdeeds: as mere vulgar criminal and disreputable persons with whom it is a disgrace for respectable persons to have any dealings, and whom all good men must, as a matter of course, look on as objects of pity not of dread, with hatred possibly, but never with fear . . . The feeling instilled into every soldier employed being, that he was altogether of a superior nature to the robber – a good man against a criminal; the plunderers being always considered not as enemies, but as malefactors.

This stern attitude was not adopted on the Punjab Frontier. There the Pathans were regarded almost as worthy enemies, wily, skilled fighters, and tough opponents. John Masters, in his autobiographical account of Indian frontier service in the 1930s, *Bugles and a Tiger*, makes it clear that even at that late date the idea persisted that frontier skirmishes were a sort of game, with the loser sacrificing life or limb, but otherwise not so very different from the team games at an English public school. Respect for the undoubted military qualities of the Pathan, at least in the spheres of individual hardiness, fieldcraft, and skill-at-arms, was allowed to develop into a sort of admiration for him as a man, and for his way of life.

The Punjab frontier was never satisfactorily settled. But the philosophical short-comings of its wardens, even if preventing them from crushing opposition from across the border, at least ensured that they were able to practise their warlike skills in a constant series of skirmishes. Indeed, as far as the military men were concerned, it was no bad thing to have a frontier where a state of virtual warfare was in existence. Training merged into real operations. Sentries had to be alert. Precautions could never be neglected. There were opportunities to show efficiency, zeal, and gallantry. There were medals, awards, and distinctions to be won, and colonels and generals could earn their knighthoods if involved in successful operations on a large enough scale.

But there is no evidence to suggest that this frontier was deliberately left unsettled in order to preserve it as a controlled training area with a real enemy. The financial burden of military defences was always too high for the Government of India's liking, and no effort was spared to reduce the estimates for frontier garrisons. The very existence of the Punjab Frontier Force as a separate organisation was due to a desire to keep the expensive regular army away from the frontier.

The existence of two separate forces side-by-side in the Punjab became a source of increasing administrative inconvenience. In the other areas where local forces were employed under the control of the Foreign Department, there were no garrisons provided from the forces of the Commander-in-Chief, India. In the Punjab, the situation was different, as large numbers of troops from the Bengal Army were stationed there. This was partly because the stations there were more healthy for European troops, partly because of the strategic importance of the province in the event of a threatened invasion from beyond the North-West Frontier, and partly because of the need to hold down a warlike and spirited population. In war, the Frontier Force was placed at the Commander-in-Chief's disposal, and came under his control for all tactical purposes, although not in matters of internal brigade or regimental procedures. In peace the situation was more complicated. In 1856, for instance, the major-general commanding the regular garrison of the Punjab desired the brigadier-general commanding the Punjab Frontier Force to alter the dispositions of a number of border posts. In order to obtain the compliance of the brigadier-general with his requirements, the major-general had to apply to the Commander-in-Chief, India, who had to address the Governor-General on the subject, who issued orders, through the Government of India in the Foreign Department, to the Lieutenant-Governor of the Punjab, that the brigadier-general commanding the Punjab Frontier Force should be instructed to conform to the major-general's wishes. Sir Charles Napier, then Commander-in-Chief, India, fulminated at the absurdity of a situation in which he, the general in supreme control of all British forces in the East Indies, could not move a single sentry at a Punjab border post.

In 1886 this situation was remedied, and the Punjab Frontier Force was brought directly under the Commander-in-Chief, India, as part of the Bengal Army. The localised character of the force was maintained, however, and it continued until 1903 to form a separate section within the Army. Its customs, traditions, and special privileges were safeguarded, and it preserved the feeling of being a unique and select body.

The Punjab Frontier Force regiments generally regarded themselves as the Indian Army's *corps d'élite*. Other Indian regiments bore titles which suggested élite status, such as Grenadiers, Rifles, and Light Infantry, but did not in fact enjoy an esteem or a role significantly different from the rest of the army. There were no Indian household troops, except for three cavalry squadrons, which were the respective bodyguards of the Governor-General, the Governor of Madras, and the Governor of Bombay. These were the senior Indian units in order of precedence, but were too small to be regarded as anything except ceremonial troops (although they performed well on active service when required). No British Household troops served in India. The Royal Horse Artillery took precedence over all other arms, but the troops which seem to have impressed Indian opinion most were the Scottish Highlanders, with their kilts and bagpipes. To such an extent were they admired as shock troops that when in the 1860s the Amir of Afghanistan raised a regular army, on the western pattern, he dressed his picked troops in kilts of his own peculiar design.

A number of plans were proposed to secure the unification of the three separate armies of Bengal, Madras, and Bombay, before this was eventually achieved in 1895. It was clear that the trend towards unification was one that could not be resisted for ever, but the long series of proposals for army reorganisation had an unsettling effect on officers and men alike. The three staff corps, to which all the officers of Indian troops in each army belonged, were combined to form one Indian Staff Corps in 1891.

The three presidency armies were abolished in 1895 and replaced by one Army of India divided into four commands. These were the Punjab (including the Punjab Frontier Force), Bengal, Madras (including Burma) and Bombay (including Sind and Baluchistan). Each was under the command of a lieutenant-general, who was, in turn, directly under the Commander-in-Chief, India. The separate commanders-in-chief and the military departments of Madras and Bombay disappeared, and their functions were divided between the lieutenant-generals of commands, the Commander-in-Chief, India, and the Military Department of the Government of India.

The usual wartime formations of the Indian Army conformed to those of conventional European warfare. Three or four regiments formed a brigade, and three or four brigades made up a division of all arms, about 10,000 strong, under the command, usually, of a major-general. Two or more divisions could be combined to form an army corps under a lieutenant-general, but in Indian campaigns, which were fought on what was, by European standards, a very small scale, a more common formation was the field force, consisting of one or two divisions only. Generally each brigade had at least one European unit, with the rest Indian.

Prior to 1889, the three main armies were divided into divisions and brigades as part of their normal peacetime organisation. The divisional and brigade commanders were responsible for the training and administration of their troops in peace, and for leading them into action in war. For reasons of internal security, however, as well as on economy grounds, the troops were scattered through a large number of different stations, usually in brigade strength, but in many cases even in single regiments. Nearly every major city in India had its own garrison. This made it difficult to assemble troops in large numbers for exercise, and the formation staffs had little opportunity of training for their active service duties.

This organisation was abandoned in 1889, and replaced by one in which each of the armies was divided into 1st or 2nd class districts under major-generals or brigadier-generals, and the districts into 1st or 2nd class stations, each under the senior officer serving there, usually a colonel. The troops continued to be split up into small groups, and the Indian Army was still deployed so that it could hold down India. From these troops, enough units were earmarked to form four active divisions, but no divisional staffs were maintained in peacetime, and it was not certain which units would be brigaded together, or who their commanders would be in time of war.

The deficiencies of this system led to a complete reorganisation of the Army in India under the direction of Lord Kitchener, Commander-in-Chief, India, from 1902 to 1909. Four major principles were adopted. These were that the army's primary role was the defence of the North-West Frontier against foreign aggression; that the organisation in which the army was administered and trained in peacetime should be the same as that in which it went to war; that the internal security role was a secondary consideration, intended merely to free the field army for other duties; and that all units should have equal training and experience of both roles.

Kitchener's first plan was not adopted in its entirety, mainly on financial grounds. The cost of abandoning some thirty-four stations and building new ones, so that troops could be concentrated in the areas of the proposed new army corps commands, was prohibitive. Instead, in 1905 a compromise was adopted. The nine divisional commands were set up, and Madras Command abolished. The Northern Command was allotted the 1st, 2nd, and 3rd Divisions, with the Kohat, Bannu and Derajat Brigades. Western Command was given the 4th, 5th, and 6th Divisions, with the Aden Brigade, and Eastern Command the 7th and 8th Divisions. The 9th (Secunderabad) Division, which was all that remained of the old Madras Army, and the Burma Division, were not allotted to a command but came directly under Army Headquarters.

Each of the divisional commands, with the exception of Burma, was organised so that on mobilisation it would put into the field a complete infantry division, a cavalry brigade (with one exception) and a number of troops who would be committed to internal security purposes or to local frontier defence. Each divisional command was under a major-general with an establishment of staff officers who formed the divisional headquarters.

Regiments were not permanently allotted to any brigade or division. To avoid troops becoming 'localised', and with the especial intention of giving every unit the opportunity of service on the North-West Frontier, the practice of the British Army was adopted, in which regiments served a tour of duty of several years with one formation and were then posted to another in a different area.

THE RENUMBERING

In order to emphasise that there was now only one Indian Army, and that its regiments would no longer be localised in peacetime in their own regions, a comprehensive scheme was worked out to renumber all Indian units in one sequence for each army. New designations were allotted to regiments, and the words Bengal, Madras, and Bombay disappeared from the Army Lists.

The renumbering was kept as simple as possible. All Bengal cavalry regiments retained their old numbers. The Madras cavalry added 25 to theirs, so that the 1st Madras Lancers became the 26th Light Cavalry. The Bombay Cavalry added 30 to their numbers, so that the 1st (Duke of Connaught's Own) Bombay Lancers became the 31st Duke of Connaught's Own Lancers. The Punjab Frontier Force Cavalry added 21 to their numbers, and various local corps, the Hyderabad Contingent Cavalry and the Central India Horse, were also brought into the common sequence.

A similar procedure was adopted in the case of the infantry. Bengal regiments retained their existing numbers, so that the 1st (Brahman) Infantry became the 1st Brahmans. The four Sikh Infantry regiments of the Punjab Frontier Force added 50 to their previous numbers. Thus the 1st Sikh Infantry became the 51st Sikhs (Frontier Force). Madras Infantry added 60, and in cases where regiments were recruited mainly from the Punjab this was recognised in their new titles. In this way the 9th Madras Infantry became the 69th Punjabis, and the 12th Burma Infantry the 72nd Punjabis. The former Hyderabad Contingent Infantry regiments were brought into the line and numbered from 94 to 99.

The Bombay Infantry added on 100, and the 1st Bombay Grenadiers became the 101st Grenadiers. The Gurkha regiments were numbered in a separate sequence from 1 to 10. The Indian Mountain Artillery batteries were numbered in sequence from 21 onwards, and the Bengal, Madras, and Bombay Sappers and Miners were numbered 1st, 2nd, and 3rd respectively.

This reorganisation did away with the last vestiges of the old presidential armies, and from then on officers and men of the Indian regiments all belonged to one service, designated the Indian Army. Kitchener has, in many histories of the Indian Army, been credited with the idea of forming the Army into divisions and brigades in peacetime. However, as we have seen, such an organisation previously had existed until 1889 when it was dismantled by Lord Roberts. Kitchener's plan, in this respect, merely put right Roberts' mistake. The originality of Kitchener's scheme lay in the proposals to move the Army's formations to peacetime locations nearer what was considered to be the most likely area of operations.

Despite the many improvements brought about by Kitchener's reforms, a number of weaknesses in the higher organisation of the Army still remained. General Headquarters, dealing directly with the divisional commanders, had to cope with many items of minor administrative detail. The lack of an intermediate chain of command resulted in excessive centralisation of duties in the Commander-in-Chief's office. The Commander-in-Chief, who in 1906 assumed the additional duties of the Army Member of the Governor-General's Council, and in 1909 combined these with the duties of Military Supply Member, was unable satisfactorily to undertake the increased burden of responsibility. The divisional commanders also had too many administrative duties to perform. They had to command and administer in peacetime a number of troops (internal security forces, volunteers, and so on) who were not part of their active divisions on mobilisation, and so could not concentrate on training their own divisions for war without neglecting their other, local, duties. On mobilisation, the divisional commander and his staff took the field with their division, leaving behind no headquarters to command these local forces, nor any machinery to ensure continuity of military administration in their normal peacetime areas. The ancillary services were under-established, and technical and administrative staff were not provided in sufficient numbers to meet the requirements of war. On financial grounds, many of the troops intended to form part of the field army or the internal security forces were not moved from their previous stations into the area of their divisional commands for either training or garrison duty.

POSTWAR DEVELOPMENTS

The reorganisation of the Army in India after World War I was intended to remedy these faults. In order to leave Army Headquarters free to concentrate on important policy matters, four commands were established, each under a general officer commanding in chief, who usually held the rank of lieutenant-general. These officers were responsible for the administration, training, efficiency and command of all forces stationed in their areas. The commands were each divided into three or four districts, under general officers commanding, usually major-generals, and further administrative authority was delegated to district headquarters. Each level of headquarters was provided with enough staff to leave behind a proportion to carry on the normal routine after mobilisation. These four commands were: Northern Command, covering the areas of the Punjab and the North-

West Frontier Province; Western Command, in Sind, Baluchistan, and Rajasthan; Eastern Command, in the United Provinces and Bengal; and Southern Command covering the Indian Peninsula.

The troops based in these commands were allotted to one of three roles, the field army, covering troops, or internal security. The field army consisted of four infantry divisions and five cavalry brigades, compared to the nine divisions and eight cavalry brigades of the prewar organisation. The covering troops consisted of twelve infantry brigades with supporting arms. Their task was to act as a frontier force to deal quickly with minor disturbances, and to act as a screen in the event of a major invasion, behind which the field army could mobilise. The internal security troops, as before, were intended to free the field army to act as a mobile striking force. As the Indian nationalist movement gathered momentum in the 1920s and 1930s these were frequently called upon to support the civil administration, and at times had to be assisted by field army units.

All troops were expected to be capable of serving in any of the three roles, but the increasing use of field army troops on internal security duties tended to interfere with their normal training for war. On the outbreak of World War II about one-third of the troops of the regular British Army were virtually locked up in India, trying to deal with nationalist discontent. Although a few Indian troops were earmarked for service in overseas campaigns, the garrison of India was at times more of a drain on British resources than a source of British strength.

From the 1920s onwards, questions of Army policy were increasingly affected by political developments within India. The Montagu-Chelmsford reforms held out to India the prospect of a steady advance towards self-government. A self-governing country would need a self-contained army, and the 1922 reorganisation of the Indian Army formed a base on which this would eventually be built. The elected Indian Legislative Assembly had the right to discuss and criticise the Government's budgetary proposals. Although military expenditure was expressly excluded from the control of the Assembly, military matters were discussed, and the members of the Assembly could register their disapproval of military policy by voting against such government proposals as were subject to their control.

In March 1921 the Assembly passed a resolution that 'the purpose of the Army in India must be held to be the defence of India against external aggression and the maintenance of internal peace and tranquillity. To the extent that it is necessary for India to maintain an Army for these purposes, its organisation, equipment, and administration should be thoroughly up-to-date.'

It was against this background of public opinion that the Indian Army began to change from the garrison of Imperial India to the defence force of a self-governing country.

The question of 'Indianising' the officer corps was regarded as a test of British sincerity in promises of eventual self-government. At the end of World War I a few cadetships had been given to Indian youths. In 1921 the Commander-in-Chief, India, Sir Henry Rawlinson, publicly supported proposals made by the Legislative Assembly to increase the opportunities for Indians to gain commissions in the Army. This alarmed the British Cabinet, which felt that the pace of progress towards majority rule should not become too fast. Rawlinson had made his proposals with a view to propitiating Indian public opinion and facilitating the passage of the Budget in the Indian Legislature. He was supported by the Governor-General, Lord Reading, but opposed by the Secretary of the Military Department in the India Office, Lieutenant-General Cobbe. Cobbe was supported by the Secretary of State for India, Lord Peel, and the attempts by the India Office to repudiate Rawlinson's proposals caused the Government of India some embarrassment.

Financial considerations were, as always, of the utmost importance. Military expenditure was unpopular, and it was proposed to reduce the British component of the Indian Army by 6,000. Rawlinson had to point out that if the number of British troops was reduced, British control over India would be weakened, and the City of London cease to subscribe to the Government of India's stocks. He also pointed out (in case he was asked to make an equivalent saving in Indian troops) that 6,000 British soldiers cost the Government of India the same as 25,000 Indian soldiers, and that to disband that number of Indian troops would alienate Indian public opinion.

No savings could be made when troops were used on internal security duties. The costs of police were normally met by the provinces in which they were employed, and, as the internal security troops were used to aid the police, it was suggested that this part of the military budget should be met by provincial rather than central Army votes. This proved impracticable, as the provinces were in debt and were borrowing money from the Government of India to pay for the services which they already maintained. In fact, provincial governments were economising on the police forces, which they did pay for, and depending for aid, in case of riots, on the Army, which they did not pay for.

Rawlinson complained that the field army was so attenuated that it could not fill its proper role. It could not think of an advance against the capital city of its most likely opponent, Afghanistan, without strong reinforcements from the United Kingdom. Its British component was at that time steadily shrinking, because the drafts from the United Kingdom needed to replace time-expired men were being diverted to deal with troubles in Ireland. A memorandum by Rawlinson in August 1922 noted that in the event of a war with the Afghans the British would be thrown on the defensive, and if any serious defeat were to happen, such as the British suffered at Maiwand in the Second Afghan War, risings within India would certainly follow.

The Indian Army, which for a hundred years had dominated the sub-continent and the Asian countries on its borders, was, at the end of the period covered by this volume, hard-pressed even to maintain its own positions.

(right) *British officer and regular troopers, full dress. 1st Madras Light Cavalry, 1845;* (below) *Indian irregular sowars, full dress. 3rd Nizam's Cavalry (Hyderabad Contingent), c 1846. Note the contrast between the European style of the regular cavalry uniform and the Indian style of the irregulars. Lithographs by J. Harris after Henry Martens*

British officer and Indian trooper in full dress. 26th and 27th Light Cavalry, 1904. Watercolour by A. C. Lovett

Indian trooper in service dress. 31st Duke of Connaught's Own Lancers, 1904. Watercolour by A. C. Lovett

3

The Fighting Arms

The Cavalry

EUROPEAN CAVALRY

The East India Company did not keep up European cavalry regiments until 1857, preferring instead to use its less expensive Indian Light Cavalry and Irregular Cavalry regiments. During the Indian Mutiny three regiments of Bengal European Light Cavalry were raised, and these, in common with the Company's other European troops, were transferred to the British Army in 1861.

The first cavalry regiment of the British Army to be sent to India arrived at the very end of the eighteenth century. Other regiments followed, although for many years only light cavalry regiments were sent to India. After the Crimean War, it was ruled that, with the exception of the Household Cavalry, the 4th and 5th Dragoon Guards, and the 1st and 2nd Dragoons, all former heavy cavalry regiments were to be considered 'medium' cavalry, and were to join the light cavalry on the roster for tours of Indian duty. In 1889 the heavy cavalry, with the exception of the Household Cavalry, also became liable for service in India. Regiments of British cavalry at this time normally spent up to nine years in India on any tour of duty there. As part of the post-Mutiny reorganisation of the Indian Army, the British cavalry component was fixed at nine regiments, about one-third of the total cavalry in the British Army. This number remained constant until the 1922 re-organisation when it was reduced to six.

The cavalry regiment was both a tactical and a social or administrative unit. Regiments were independent of each other, and had no ties with other regular or reserve units. Their uniforms indicated to which branch of the cavalry they belonged – dragoon guards, dragoons, hussars, lancers, or light dragoons – but in other matters the regiments had little common ground.

Organisation varied from time to time, but in general a British cavalry regiment was some five or six hundred strong, commanded by a colonel or lieutenant-colonel with a small regimental headquarters. In war, regiments were divided into squadrons, made up of two troops each, but not until the 1880s did this become part of the normal peacetime organisation. Prior to then, cavalry regiments were divided into troops for all administrative purposes. Each troop was commanded by a captain with two subalterns, and consisted of about eighty men. In the twentieth century a troop of cavalry was reduced both in numbers and status to become a subaltern's command.

Cavalry were trained primarily for shock action. This was especially the case in India, where the European was heavier, man and horse, than his likely Indian opponents, and where the conventional light cavalry duties of scouting, patrols, and escorts could be more efficiently carried out by Indian irregular horse. The sword was their main weapon, straight for the heavy cavalry, slightly curved for the light horsemen. Lancer regiments carried the weapon from which they took their name, in addition to the light pattern sword. Carbines were carried by all regiments from the 1860s on, until replaced by a short magazine rifle at the beginning of the twentieth century. Dismounted work was unpopular, however, and musketry training was neglected by both officers and men. The resemblance between the red-and-white marker flag used to signal 'missed' at the butts, and the red-and-white pennon of a cavalry lance, was frequently remarked on by the rest of the Army.

INDIAN CAVALRY

The first three regiments of native cavalry to be trained in the European style were raised by the Company in Bengal, in 1796. Further regiments were formed, and in 1825 their title was changed to light cavalry. By 1857 the Bengal Army had ten regiments of light cavalry, the Madras Army eight, and the Bombay Army three.

Despite their title, these regiments, like the British 'light' cavalry in India, were light only by courtesy. They certainly dressed in the style of European light regiments, except that their jackets were 'French gray' (a sort of light blue) instead of the dark blue or scarlet of the British coatees, but in training, equipment and employment they were assimilated as closely as possible to the European dragoons whose role, as mounted shock troops, they were intended to share.

These regular light cavalry regiments consisted of 24 European officers and about 400 men. Each of the six troops was commanded by a captain, assisted by a lieutenant. There were also an adjutant and a quartermaster-interpreter (both senior lieutenants) and four cornets, with a surgeon and veterinary surgeon at regimental headquarters. Each troop was made up of a subedar (Indian captain), a jemadar (Indian lieutenant), four havildars (Indian sergeants), four naiks (Indian corporals), a trumpeter, a farrier, and about sixty private sepoys. Each horse was allotted its own grass-cutter and a one-third share of a groom. Officers' and NCOs' horses had one groom each. The men carried a European light dragoon sword, and pistols. One man in each section of four carried a carbine. As with the British cavalry, troops were paired in battle to form squadrons but remained the normal sub-unit for administrative purposes.

The true light cavalry work was at first performed by horsemen of the Company's Indian allies, armed and raised in the local fashion, each man providing his own horse and equipment. As the number of allies capable of providing such troops dwindled, the Company began to raise such units as part of its own army, under the title 'local horse'. In the Bengal Army these were renamed, in 1840, irregular cavalry, and by 1857 consisted of eighteen regiments numbered in their own separate sequence. The Bombay Army raised seven regiments of 'irregular horse' each bearing the name of the area where it was localised. The Madras Army did not need to raise irregular cavalry, as its Indian campaigns virtually came to an end with the defeat of Tipu in 1799. It could, however, count on the services of the four cavalry regiments of the Nizam of Hyderabad's Contingent, all of which were formed on the irregular principle.

The twin characteristics of the irregular units were that they required only three or four European officers in each regiment, and that the men were armed, dressed, and maintained in the local Indian style rather than in imitation of the European.

The four officers were the commandant, second-in-command, adjutant and surgeon. These had the satisfaction of holding more interesting and responsible posts (and better paid ones) than they could have hoped for in the regular regiments from which they were drawn, while the Indian officers, ressaldars or squadron commanders, were also able to enjoy more authority and status. Although the European officers enjoyed the extra pay of their 'regimental staff' appointments, their basic pay was not an additional charge against the military budget, since they continued to be held on the lists of the regular regiments from which they were drawn, and the vacancies there were not filled up.

In their uniforms, the irregulars discarded the unsuitable European style of dress for the more practical costume that had for centuries been worn in India. The tall cylindrical shako was replaced by the *pagri* or turban. The padded and quilted jacket gave way to the long loose *kurta* or frock coat. Narrow overalls, tightly strapped under the instep, were discarded in favour of riding breeches, worn either with puttees or with long boots. Sharp, Indian pattern swords, the talwar, or the scimitar-like shamshir took the place of the light dragoon sword which Indian troopers found unwieldy and cumbersome. The contemporary dragoon saddle, which pushed the rider so high above his horse that he virtually rode by balance alone, and which was most suited to a man so short in the body and so long in the leg as to make him, in the words of one critic, 'look like the afternoon shadow of somebody else', was abandoned, and local pattern saddlery substituted.

Irregular cavalry was organised on what was known as the silladar principle. The silladar was a man who, in return for a higher rate of pay than that of the ordinary regular, contracted to provide and maintain his own horse and equipment. This meant that the Government was not obliged to find rations for either the man or his horse, nor to look after them while sick, nor to provide transport for them in the field. This cut down on the need for ordnance depots, supply trains, commissariat, and similar organisations forming the 'tail' of a civilised army.

To ensure uniformity, the regiments themselves acted as contractors, supplying recruits on enlistment with a horse, equipment, and clothing of regimental pattern. The recruit paid to the regiment a sum corresponding to the value of what he was given, called an assami, which can be translated as his 'berth', or 'billet'. On being honourably discharged or invalided a man's assami was refunded to him, or he could choose to keep his horse and equipment instead. If a man was dishonourably discharged, he forfeited his assami, which thus acted as a bond for good behaviour. The assamis of men who died were refunded to their relatives. If a horse was unfit for duty, its rider's pay was reduced to that of a foot soldier until he could find another mount. Replacements for both horses and equipment had to be paid for by the silladar, so that he had an incentive to take care of them. Inevitably horses and equipment had to be replaced at some time, despite even the greatest care, but to avoid such heavy expenses being incurred all at once, the silladar's pay was subject to monthly 'cuttings' which enabled him to build up a reserve sum for this purpose. Horses killed or drowned on active service were replaced at the expense of a regimental fund to which all silladars subscribed equally. Only firearms and ammunition were issued free by the Government.

When the Indian Army was reorganised after the Mutiny, the irregular rather than the old regular regiments were taken as a model. The superiority of the former over the latter was by no means clear. Certainly all ten Bengal regular cavalry units had mutinied, but so had more than half of the Bengal irregular cavalry, while the regular light cavalry of Madras and Bombay had remained loyal. The economic advantages to the Government of the silladar system could not be denied, however, and by 1903 only three out of the thirty-nine regiments of Indian cavalry were non-silladar. The greater comfort of Indian-style costume was also admitted, and even the non-silladar regiments, all of the Madras cavalry, gradually discarded their European-style uniforms in favour of more practical Indian ones.

In the establishments of units a compromise was effected between the old irregular and regular systems. Regiments were commanded by a colonel with an adjutant and small headquarters staff, and divided into three squadrons (increased in 1885 to four squadrons) each with a British squadron commander and second-in-command. Each squadron was divided into two troops, and each troop was commanded by a ressaldar assisted by a jemadar. The rank and file of each troop consisted of a kot-dafadar and eight dafadars (the irregular cavalry term for a sergeant, or corporal-of-horse) with seventy sowars (troopers) and a trumpeter.

The standard armament continued to be the sword, usually of a curved Indian pattern, with a sharpened edge used in preference to the point. (British cavalrymen were trained to use their swords primarily with a thrusting action rather than a cutting one.) Just over half the Indian cavalry were designated lancers and equipped with the lance accordingly. Carbines were issued to all cavalry troopers from about 1860 onwards, and these were replaced by short rifles in the early 1900s. Dismounted work was even less popular with the Indian cavalry than with the British. Their officers prided themselves on commanding the finest irregular cavalry in the world, and opposed attempts to make their regiments anything other than that. The appointment of an Inspector General of Cavalry after the Second Afghan War, to bring some attempt at standardisation, was greeted with horror, and cries that they were 'being dragooned'. Sufficient resistance to orthodox methods was put up to lead one British fusilier officer in India to call his cat 'Indian Cavalry' on the grounds that when he wasn't eating or sleeping he was playing games or making love. Despite this jaundiced view, the Indian horsemen, under the British, maintained their traditional reputation for skill-at-arms, dash, and equitation. When properly led and commanded, they proved more than a match for any light cavalry ever put into the field against them, and in their picturesque uniforms, gaily coloured turbans, and coats of dark or light blue, green, scarlet, or yellow, they made an impressive appearance to friend and enemy alike.

The silladari system worked quite well as long as the Indian cavalry was required only for garrison duties or relatively small wars on the borders of India, but it was unsuited to large-scale modern warfare. It broke down during World War I for several reasons. Resupply of regimental-pattern equipment proved impossible in overseas theatres. Regiments lacked the liquidity to refund the assamis of unprecedentedly high numbers of men killed or wounded on active service. Accounting problems at the under-manned depots proved almost insuperable. It was clear that the levying of horsemen by private contract was an anachronism. In the 1922 reorganisation, the number of cavalry regiments was reduced from thirty-nine to twenty, all equipped and maintained by the Government as in other modern armies. These regiments were divided into a headquarters and three sabre squadrons, each subdivided into three sabre troops and a light machine-gun troop.

The Artillery

THE EAST INDIA COMPANY'S ARTILLERY

With the exception of a few companies of the Royal Artillery which served in India during the 1750s, all artillerymen in the Indian Army were in the Company's service until the crisis of the Indian Mutiny. In 1861 the Europeans were transferred to the Royal Regiment, and most of the Indian personnel were disbanded.

The European artillery had the best men the Company could find. Their officers were, from the beginning of the nineteenth century, appointed from cadets trained either at the Royal Military Academy Woolwich, where all future officers of the Royal Artillery and Engineers were educated, or at the Company's own Military Seminary, Addiscombe. The ordinary soldiers were the pick of the Company's European recruits, and in view of the tactical importance of artillery, if men were in short supply, priority was given to the needs of that arm.

There were three regiments of artillery, one for each of the three presidential armies. The usual unit was the company of a hundred or so men, under a captain and his lieutenants. This unit included all the gunners, artificers and tradesmen needed to man a battery of six field guns, or a varying number of siege or garrison pieces. In 1820 the growing number of companies were grouped into battalions. These battalions, like those of the Royal Regiment of Artillery, were administrative organisations, not battle formations. They served to provide a lower-level headquarters for the companies of which they were formed, and to provide a better career structure for the officers by creating a greater number of senior posts and staff appointments.

By 1857 there were six European battalions in the Bengal Artillery, four in the Madras Artillery, and three in the Bombay Artillery, each battalion being made up of four companies. They wore dark blue coats with scarlet facings, and were virtually indistinguishable in their uniforms from the Royal Artillery.

Each regiment had its own élite corps of Horse Artillery, following the example of the Royal Horse Artillery. As with the RHA, officers for the Indian horse gunners were selected from the dismounted branch, and then served alternate tours in each branch as they were promoted. The dress of these troops was, if anything, even more splendid than those of the British service, for although both wore a short blue jacket decorated with rows of gold lace and ball buttons, the Company's men wore, instead of the RHA hussar busby, a great Roman helmet, like that of the French cuirassier, with a long flowing mane of red or black horsehair. Formed in the early years of the nineteenth century, there were by 1857 three brigades in the Bengal Army, each composed of three European and one native troop. (The first brigade had a fifth, native, troop, a remnant of the Shah Shuja's Contingent raised by the British during the First Afghan War.) The Madras Horse Artillery was composed of four European and two native troops, and the Bombay of four European troops only.

Each troop corresponded to a company in that it provided the personnel to work a battery of six guns. Unlike the companies, the troops had a permanent complement of drivers, since they were horsed at all times, and not used to man-fixed defences, although at sieges any horse artillerymen in the investing force might take a turn in the siege batteries. Horse artillery brigades corresponded to battalions of foot artillery.

When the Indian artilleries were brought on the British establishment in 1861, the

Royal Artillery, which had previously totalled one horse brigade, five field brigades, and ten garrison brigades, was increased by the addition of four horse brigades, four field brigades, and six garrison or mixed brigades, all of five batteries each.

Immediately prior to amalgamation, the Royal Regiment had changed its nomenclature. All troops and companies were renamed *batteries*, a term which had hitherto referred to the actual guns rather than the men who served them. The battalions were renamed *brigades*, which remained administrative units. It was not for another forty years that the brigade of several batteries under a lieutenant-colonel became a separate tactical unit, corresponding to a cavalry regiment or infantry battalion.

Artillery units composed of Indian personnel, known as 'Golandaz', a local term meaning 'ball-throwers', were raised in the eighteenth century, but these were at first allotted to the European companies for unskilled work, and for political reasons were not actually allowed to work or fire the guns. Each battery normally had a company of European gunners and a company of gun lascars when fully manned for service.

Early in the nineteenth century, however, companies of Indian gunners were formed in each of the three armies, as part of the existing regiments of artillery. These were formed into battalions as their numbers increased, and by 1857 there were three Indian battalions in the Bengal Army, one in the Madras Army and two in the Bombay Army. Each battalion was then made up of six companies. The Bengal and Madras Artilleries also raised horse troops manned by Indian gunners. These were distributed among the horse brigades as described above.

The record in the Indian Mutiny of the native artillerymen was a little better than that of the other arms. Two out of four horse troops and six out of eighteen foot companies of the Bengal Artillery joined the mutineers or were disbanded, and two of the Bombay companies mutinied. Unfortunately for the future of their arm, the mutinous gunners served their guns with considerable skill and effectiveness, and the result was that in the post-Mutiny reorganisation nearly all Indian artillery was disbanded on security grounds. The European gunners were transferred to the Royal Artillery. The Indian personnel were gradually reduced. Some transferred to Indian cavalry or infantry regiments. Others took their discharge or were pensioned off. Some lingered on for a time attached to British artillery batteries in the same way as the old gun lascars, or were allotted to the fixed defences and coast artillery.

The officers of the former Indian Artilleries were given commissions in the Royal Artillery. But they were kept on separate lists for promotion, so that no officer of any of the four regiments was superseded or held back from the promotion (by seniority) that he would have been entitled to had the amalgamation not occurred. This system was also adopted in the case of the Royal and Indian Engineers.

THE INDIAN MOUNTAIN ARTILLERY

The only Indian batteries remaining as fighting units were three light field batteries, a garrison battery, and two mountain batteries in the Punjab Frontier Force; one mountain battery in the Sind Frontier Force; and four light field batteries in the Hyderabad Contingent. In 1876 the light batteries were converted to mountain batteries, and the 1st Company of Bombay Native Artillery was re-equipped as a mountain battery. As part of the reorganisation following the Second Afghan War, the Punjab mountain batteries were reduced and numbered one to four, the two Bombay batteries were allotted the numbers five and six, and two new batteries were raised as part of the Bengal Army, to be given the numbers seven and eight.

As part of the Kitchener reforms the Hyderabad batteries disappeared, but another four Indian mountain batteries were raised, making a total of twelve, numbered 21st to 32nd. During World War I another fifteen batteries were formed, and several of these remained in being after the 1922 reforms. This left the Indian Army with twenty-one pack batteries of what was then the Royal Indian Mountain Artillery, numbered 101st to 121st. In all these batteries, whether mountain or frontier garrison, the officers and any attached European personnel were members of the Royal Regiment of Artillery in the British Army, and did not belong to the Indian service. They served a tour of duty with the mountain batteries, when their salaries were paid for by the Government of India, plus extra allowances, in the same way as other British Army personnel stationed in India. There was no special corps of officers for the Indian artillery, although officers could continue in that branch of the Regiment if they wished.

Because of the limited number of these batteries, any would-be Indian artilleryman had a great deal of competition to face, and only the highest standard of man was recruited. In the same way, Royal Artillery officers vied with each other to fill the small number of vacancies (in 1903 there were 47 officers and 3,000 Indian artillerymen). Thus the mountain batteries became something of a *corps d'élite*, especially because there was a greater chance of active service (with its chance of professional advancement and awards), with them in hilly frontier terrain where wheeled artillery could not go.

Indeed, this was officially recognised in 1902, when the Royal Artillery was divided into two distinct branches, the Royal Field Artillery and the Royal Garrison Artillery. The Royal Horse Artillery had since its formation been the *corps d'élite* of the whole Regiment, with picked officers serving with the horse batteries and the line in alternate tours. It was now made the *corps d'élite* of the mounted or Field branch, and the Mountain Artillery, British or Indian, was given the same role in respect of the dismounted branch, the Royal Garrison Artillery. Although on first appearance there might be little in common between the small mountain guns and the large siege or fortress guns manned by the rest of the Garrison artillery, the point of similarity was that both types of gun were served by dismounted gunners, men who walked into battle behind the horse, mule, or bullock gun teams, rather than rode into battle on the gun-limbers or horses.

The field artillery, originally foot artillery, had become mounted after the Crimean War, and had derived increased status on that account, despite the jibes of the RHA about the foot artillerymen calling their trousers 'overalls' and riding on cart horses. The garrison and siege batteries had remained dismounted and were therefore considered less dashing, and less fashionable. Despite the greater scientific and technical ability required for this branch, it was one which few officers joined willingly.

THE ROYAL ARTILLERY IN INDIA

Following the disbandment of the Indian artillery and the absorption of the Company's European artillery, the Royal Artillery had to fill the gap in the Indian Army's order of battle. Batteries were sent to India on tours of duty of about thirteen years at a time. Instead of the twenty-three troops of horse artillery and ninety-two field and garrison companies that the East India Company's Artillery maintained, the Royal Artillery provided, in 1864, eighty-six batteries, horse, field, and garrison. This number remained much the same for the rest of the century, but was increased slightly after the Kitchener reforms. The 1906 establishment shows a total of ninety-two British units, composed of eleven horse batteries, forty-five field batteries, six heavy batteries, eight mountain batteries (not including the Indian batteries) and twenty-two companies of Royal Garrison Artillery (the RGA having adopted the old title of 'company' to indicate the men rather than the variable number of guns they were to serve).

The gun line of the batteries was organised in three 'divisions', each of two guns commanded by a subaltern. In India, batteries consisted, on average, of about 150 British artillerymen, including gunners, drivers, and tradesmen of various sorts. In 1874, battery commanders were upgraded from the rank of captain to that of major, and the old artillery rank of second captain was replaced by the appointment of 'battery captain', or captain who was second-in-command of the battery. Batteries with bullocks or mules were allotted 100 or 200 Indian personnel respectively, as drivers and grass-cutters. By World War I most artillery was organised in brigades of two or three batteries, with a headquarters under a lieutenant-colonel, and supporting ammunition columns.

Ammunition columns were Royal Artillery units and formed an integral part of the supply system in European campaigns; they were responsible for bringing ammunition for both artillery and small arms from the ordnance parks to the rest of the army. The ordnance parks were usually one day's march behind the fighting line, and the fighting units thus had only to send their wagons to their nearest supporting artillery's ammunition column. This arrangement avoided large numbers of vehicles clogging the roads in rear of the army. For reasons of economy, the British Army did not keep up ammunition columns in peacetime until early in the twentieth century, and, prior to that, formed the ammunition columns required for batteries of an expeditionary force by simply taking the men and horses from the batteries remaining at home. In India, no ammunition columns existed at all until 1904, and until then the Ordnance Field Park marched with the fighting formations which it supplied.

The reforms of 1922 reduced the effective strength of artillery in India by the equivalent of five or six batteries, although the experimental use of mechanical artillery tractors (heavy motor lorries) for the heavier calibres went some way to improve mobility.

Various types of personal weapon were carried by artillerymen at different times, and according to the duties each man performed. Officers of foot artillery carried the straight, infantry officers' pattern sword until about 1859, when all artillery officers adopted the light dragoon pattern with a slightly curved blade which had prior to that time been used only by the Horse Artillery. At the same time gunners of field artillery discontinued the sword-bayonet which had been their personal arm in favour of the long sword of the horse gunners. No personal firearm was carried in the horse and field batteries, except for two muskets or carbines on each gun limber, to be used by sentries, or for local defence when guns could not be brought to bear. It was argued that in an emergency the gunners should be working their guns, and should rely on them for protection. A similar argument was used in respect of the drivers, who were issued with no arms at all, on the grounds that the drivers, in a crisis, would more than ever be concerned with the management of their gun teams, and they would be distracted from this essential duty if they were using personal weapons against the enemy.

This argument, while tenable in 'civilised' warfare, or set-piece European-style battles, was quite invalid in the campaigns against wild tribesmen on the frontiers of India. Hostilities did not cease when the guns came out of action. On the contrary, it was when they were in bivouac, or on the line of march, that attack was most likely, and there were several instances when, because of the inconvenience of large numbers of soldiers being effectively unarmed, it became necessary to issue carbines to the gunners and tradesmen, and revolvers to sergeants and drivers.

The garrison artillerymen, although less exposed to unexpected attack, were generally armed with a light musket, or carbine, and bayonet, for use in defending their works if the enemy effected a lodgement. Siege guns were helpless if attacked by an enemy on foot, and so their detachments were equipped in the same way, with a personal firearm for each man.

At the beginning of the twentieth century, swords were given up in the Field Artillery, except by officers. The Royal Garrison Artillery was re-armed with rifles and bayonets, and the Royal Field Artillery was issued with revolvers and short rifles to enable each man to have his own personal firearm.

The equipment of artillery in India was usually obsolete by European standards. Royal Artillery batteries arriving in India found themselves serving pieces that had long gone out of use at home. This naturally had a discouraging influence on their gunnery. They tended instead to take advantage of the favourable conditions of the country for the upkeep of horses for their teams, and for exercising the batteries in open country. Thus it came about that while the Artillery in the United Kingdom could shoot but not ride, that in India could ride but not shoot. Table 1 shows the equipments generally in use at different periods.

The technical advances in artillery equipment in the second half of the nineteenth century corresponded with the increasing industrialisation and scientific progress of European society. The table gives a very approximate indication of the main types of armament with batteries in India during the period under review, although, with various rearmaments, there were periods when equipments of different categories were in use simultaneously. The maximum effective range of the lighter pieces gradually increased from the 1,000 yards of the 9-pounder smoothbores to the 8,000 yards of the 18-pounder quick-firers. The rate of fire varied tremendously from equipment to equipment, depending upon the state of training of the detachments, the nature of the ammunition ordered, and other variables. The horse artillery, with smoothbores, could average about two rounds per minute. The adoption of the quick-firing system, in which a buffer and recuperator mechanism absorbed the shock of recoil and did away with the tiring business of 'running-out' the guns by hand (ie, pushing them back to their original firing platforms after every round), did much to speed up the rates of fire at the beginning of the twentieth century.

Until then guns were normally laid by the direct method, that is the Nos. 1 of the gun detachments (sergeants) aimed them at targets directly in sight from the gun position. The officers were responsible for deploying their guns, target selection, and fire control, observing the fall of shot, and ordering corrections as necessary. In Indian campaigns by far the most usual tactical unit was the division (later called a section) of two guns.

A variety of animals were used to provide mobility for the guns. The horse artillery, designed to keep up with the cavalry, was almost always drawn by teams of six horses. Field artillery in India was at first normally drawn by bullocks, but horse teams came increasingly to be used in the 1850s, to provide greater mobility in battle, and from the 1860s were used as the conventional method. The heavy guns, until the early years of the

Table 1

Artillery equipments in India, 1822–1922

Period	*Category*	*Horse*	*Field*	*Heavy*	*Mountain*
Before 1865	Smoothbore, muzzle-loading	6-pdr gun and 12-pdr howitzer	9-pdr gun and 24-pdr howitzer	18-pdr gun and 8-inch mortar	3-pdr gun and 12-pdr howitzer
1865–77	Rifled, breech-loading	9-pdr gun	12-pdr gun	40-pdr gun and 8-inch mortar	6-pdr gun
1877–85	Rifled, muzzle-loading	9-pdr gun	9-pdr gun	40-pdr gun and 6·3-inch howitzer	7-pdr gun
1885–1905	Rifled, breech-loading	12-pdr gun	12-pdr gun	40-pdr gun and 6·3-inch howitzer	2·5-inch gun (rifled, muzzle-loading)
After 1905	Rifled, breech-loading (quick-firing)	13-pdr gun	18-pdr gun	30-pdr gun and 5·4-inch howitzer	10-pdr gun

twentieth century, were pulled by elephants, with spare teams of bullocks. When action was likely, the practice was to change from elephant to bullock draught, since the elephants were unreliable under fire. At best they might stop and refuse to go forward. At worst they might panic and rush off, trampling friend and foe alike. Under the Kitchener reforms these, too, were replaced, and teams of heavy draught horses were used instead. The mountain guns were dismantled and carried in pieces on pack mules. The most famous of these equipments was the 2·5-inch jointed gun, the 'screw-gun', so called because the piece was made up of two sections, screwed together when the gun was brought into action.

In all cases alternative means were used from time to time. Bullock teams were allocated to the field batteries as late as the 1880s, on long marches, to relieve the gun horses. However, battery captains (who were in charge of the transport and wagon lines) too often used them alone on long marches, to keep their horses fresh for battle, and this frequently resulted in the oxen foundering from exhaustion through being driven at too fast a pace, and so leaving the horses to pull the guns without a relief. Far too many native bullock drivers were allowed to adopt the cruel habit of twisting their beasts' tails, which caused a great deal of suffering and lasting damage. Elephants, too, could be used to help the lighter natures of ordnance, either in draught, or by pushing them in difficult places, or by carrying them on their backs. Mountain guns could be fitted for draught and their mules used as teams in suitable conditions. Camels were used at times both to carry light guns in pack and to pull them, and were specially suitable in desert tracts from their ability to go without water for days at a time. In jungles or very broken country the mountain guns and light mortars were sometimes man-packed, carried by teams of locally recruited porters.

Engineers

THE CORPS OF ENGINEERS

The three Indian armies raised their own separate corps of Bengal, Madras and Bombay engineers respectively in the 1770s. As was then the case in the Royal Army, these engineers were staff officers and technical advisers rather than executive commanders. The uniform of the four corps corresponded in all major points, with the officers wearing a red coat with blue facings.

As part of the post-Mutiny reorganisation, the Indian Engineers were amalgamated with the Corps of Royal Engineers in 1862. The existing officers were granted royal commissions, but remained on separate presidency lists for promotion and seniority purposes. New officers were appointed to the enlarged Corps of Royal Engineers and went to India for a tour of duty either as staff officers or in command of sappers and miners. Any officer wishing to spend his whole subsequent career in India could do so if he wished, and after eighteen years' service in India could retire from the Army with an Indian Army pension. However, if an engineer officer wished to return to general corps duties in the United Kingdom or elsewhere, he could do so on six months' notice.

No companies of Royal Engineers or Sappers served in India, except briefly during the crisis of 1857–8, but to supply the European NCOs who continued to be required with the Indian Sappers and Miners, three RE companies were formed in India and numbered the 41st (for Bengal), 42nd (for Madras), and 43rd (for Bombay). These came into being in 1868, and all Royal Engineers of the rank and file serving in India were borne on the strength of the company for the presidency army in which they were working. In addition two extra companies were formed at the RE Depot, Chatham, where the East India Company's European troops had once been trained, for the purpose of training and supplying British sapper personnel for India.

THE SAPPERS AND MINERS

There were three Indian corps of sappers and miners. The words 'sap' and 'mine' relate to siege techniques, but, unlike the Indian Engineers, who were primarily 'fortress' rather than 'field' engineers, the Indian Sappers and Miners evolved out of the need to enable the Army to move in a country singularly ill-provided with roads and bridges. Indeed, this remained their prime task until late in the nineteenth century. During the Second Afghan War the Bengal companies had only five artificers each (two carpenters, two smiths, and a mason) all of whom were civilians, while the military personnel were trained only in digging, spar-bridging, and pontooning.

The Bengal Sappers and Miners originated in the Bengal Pioneer Corps raised in 1803. In 1808 this became the Corps of Pioneers and Sappers, organised in eight companies of ninety men each, with a company of miners. Two British sergeants were allotted to each company. The miners were under an engineer officer and the rest were commanded by infantrymen seconded from their own regiments. In 1819 two companies and details from the rest formed the new Bengal Sappers and Miners, and the remaining pioneers were absorbed into that corps in the reduction of 1833. In 1847 it was divided into three companies of sappers and miners, under engineer officers, and seven of pioneers, under officers from the infantry, and changed its title to the Bengal Sappers and Pioneers. The pioneers received lower rates of pay than the sappers, and wore the dark green uniform of the earlier Pioneer Corps although the sappers continued to wear the scarlet coats adopted in 1819. Sapper vacancies were filled by selection from the pioneers as they arose. The pioneer companies were numbered separately, and had their own rolls for promotion, but remained part of the same corps as the sappers, under an Engineer Commandant. In 1851 the division between the two branches ceased, and the title Bengal Sappers and Miners was again adopted. The number of companies was raised to twelve, and it was planned to replace the infantry officers by engineers as vacancies occurred.

In 1857 about 60 per cent of the Bengal Sappers and Miners went over to the mutineers. The remainder, about 500 in number, remained loyal, and three small companies, of about 40 men each, played a notable part in the Siege of Delhi. The 'Powder Bag Party' which blew in the Kashmir Gate prior to the final assault included fourteen Bengal Sappers and Miners in its number.

No great change was made in the establishment of the Bengal Sappers and Miners after the Mutiny. The number of companies was reduced from twelve to ten in 1863. The remaining infantry officers were replaced by Royal Engineers (although one infantryman remained in post as late as 1875).

The Madras Corps had its origins in two companies of pioneers raised in 1780. By 1813 these had grown to two full battalions, each of eight companies. A decision to raise a separate sapper and miner corps, made in 1818, was rescinded three years later, and the pioneers continued to do all the field engineering of the Madras Army until 1831, when the 1st Battalion became the Corps of Madras Sappers and Miners, under officers from the Madras Engineers. The 2nd Battalion was absorbed into this in 1834 as an economy measure.

The Quartermaster-General of Madras opposed this scheme since, while pioneers came under his command, sappers were controlled by the Chief Engineer. The pioneer officers, he said, might not have studied engineering, but the best roadmaker in England (McAdam) was a former ship's purser. Engineers were ignorant of the way in which troops manoeuvred, and objected to obtaining that knowledge. 'Considering their own as the first of all arts, they look down with disdain on every other branch of the service.'

However, it was fair to say that the pioneers had been doing the duties of sappers, and that for most tasks simply requiring labour it was cheaper to hire civilians than to maintain, at least in peacetime, a separate military labour corps.

By 1922 the sappers and miners were virtually the only troops still recruited in the Madras presidency. The men of this corps were among the most intelligent of all the Indian Army, despite being recruited from classes despised by the rest of the Army as 'non-martial'. In 1855 the Governor of Madras addressed the men of the corps, praising their achievements and speaking in the Urdu language, the *lingua franca* of the Indian Army, but an Indo-Germanic language from the courts of northern India. On learning that the Madrassis, who spoke a Dravidian tongue, had not understood him, he asked the senior Indian officer to translate his speech into a language that they could understand. The subedar-major accordingly repeated the speech to the men in English.

This corps was generally ahead of the other two in its standard of training. It was the first to abandon the battalion system in favour of independent companies, and as early as 1838 it was decided that four of the six companies into which it was then divided should work with the Revenue Board on civil engineering, while the other two were trained at their depot, Bangalore, in field and fortress engineering. Companies changed round every two years, to ensure that a proper cycle of training was kept up, and companies were allotted a full scale of tools, axes, crowbars, bill-hooks, hammers, and digging implements, to render them capable of immediate independent action.

The Bombay Corps, too, had its origin as pioneers. By 1822 the Bombay Pioneers consisted of one battalion of eight companies. In peacetime they were employed in making and cleaning roads in difficult areas, in the construction and repair of fortresses, in the demolition of dismantled native forts, in assisting inside magazines, and in building work at military stations. In war they were used on military roads, and at sieges, digging, tunnelling, and constructing field fortifications. A separate company of Bombay Sappers and Miners was formed in 1820 out of the existing bridging train, and a second company was formed, under Bombay Engineer officers, in 1826. These two companies became part of the Bombay Engineers in 1829, and in 1830 the Bombay Pioneers were absorbed into this single engineer corps. In 1837 the title of Bombay Sappers and Miners was revived, and infantry officers were again posted to the companies, as there were not enough engineers to officer them. But during the 1870s infantry officers were gradually replaced by Royal Engineers. Companies were commanded by captains, each with a lieutenant as second-in-command and two or three European warrant officers or sergeants as assistants. The Indian personnel of each company comprised a subedar, a jemadar, and about 150 NCOs and sappers.

Since the sappers were required to fight in their own defence if attacked, personal weapons were carried by all ranks. Up to about 1845 the standard firearm was the 'fusil', a shorter and lighter version of the ordinary 'Brown Bess' or Tower musket. The Bombay corps was then re-armed with the Brunswick rifle, and the other two with what was called a 'sapper carbine', although actually it was considerably longer in the barrel than the weapons to which the term carbine is usually given. These were replaced by the Lancaster rifle in 1866, as carried by the Royal Engineers, and this, in turn, was replaced by the Snider carbine in 1875. From 1888 the standard infantry pattern rifle was carried.

'Submarine mining' or the laying of mines in the sea in order to protect harbour entrances, was, until shortly before the outbreak of World War I, a military rather than a naval responsibility. Mines for the defence of Calcutta and Rangoon were authorised in 1868, but no major action was taken until 1879 when two companies, one each from the Madras and Bombay Sappers and Miners, were given instruction on mine-laying by a team of RE instructors sent from England with the most recent equipment. The companies were almost immediately sent on active service in the landlocked territory of the Amir of Afghanistan. Various corps of Indian submarine miners were formed subsequently, under RE officers, until the mining of Indian ports by the Army was discontinued in 1907.

PIONEERS

The Indian Army possessed a number of pioneer battalions, capable of performing simple field engineering tasks, and also of taking their place in the line of battle as infantry. The original pioneer units had been converted to sappers and miners mainly on the grounds of economy, in that it was cheaper to hire local civilian labour for the construction of roads and earthworks when required than to keep up regiments of trained soldiers for this purpose. During the Indian Mutiny, pioneers were recruited in the Punjab for service in areas where the local civilians were no longer to be trusted, or where the normal economic processes had failed to attract sufficient labour. As part of the post-Mutiny reforms, some of these Punjab pioneers were brought into the depleted line of Bengal Infantry, becoming the 23rd and 32nd (Punjab) Regiments, but retaining the designation and role of pioneers in addition to their new employment as infantry battalions. This combination proved particularly successful during the Second Afghan War, when the Indian Army was once more operating in a hostile countryside. Coolies and labourers could be recruited or imported only with difficulty, and when employed were defenceless against sudden enemy raids. The pioneer regiments were able to cope with this problem, and Sir Frederick Roberts, in compiling the force with which he made his famous march to the relief of Kandahar in 1880, left behind his trained sappers and miners, relying instead on the 23rd Pioneers to cope with any elementary field engineering and to swell his firing line when engaged with the enemy.

Following the success of these two units, other pioneer regiments were raised, or formed by the conversion of existing ordinary infantry battalions. In the 1903 order of battle, one was allotted to each of the nine divisions of the field army, one to the Army headquarters, and two to the internal security or frontier covering force.

Kitchener, a former Royal Engineer, was an advocate of pioneer battalions and it was mainly through his influence, as Secretary of State for War at the beginning of World War I, that the British Army, which had not had separate pioneer units since the seventeenth century, again brought them into its order of battle. Some battalions of infantry regiments were converted to divisional pioneers, and new labour battalions were formed. However, at the end of that war all were disbanded, and it was not until World War II that the present Royal Pioneer Corps was formed as a permanent addition to the British Army. Pioneer units were a regular part of the large European armies, but the British had never the men or money to spare for this specialist role, and relied, for labour, upon working parties, mainly provided from the infantry, under engineer supervision. Each infantry battalion had a pioneer sergeant with ten pioneers capable, in barracks, of performing elementary carpentry or plumbing tasks, and, in the field, of route clearing, obstacle removal, and the construction of field defences.

Unlike the sappers and miners, whose tools were carried on mules or in the backs of wagons, the Indian pioneers carried their tools on their backs in special leather equipment, and so differed in outline appearance from all other troops. Each sepoy carried either a pickaxe or a light spade. NCOs and buglers carried axes, saws and bill-hooks. Heavier equipment, shovels, explosives, crowbars and so on, were carried on pack-mules.

In the later stages of World War I, the pioneers were considered too valuable to be used as assault infantry, and they were used almost exclusively on pioneering work. The 1922 reorganisation recognised their changing role, and they were taken out of the infantry line altogether and organised separately as three regiments each of four battalions. The battalion organisation was subsequently dispensed with, and the pioneer companies became independent units, freed entirely from the burden of an infantry role. In 1932 the pioneers were disbanded and their duties handed over to the sappers and miners. During World War II, the Indian Pioneer Corps was raised, to supply skilled labour under military control.

The Infantry

EUROPEAN INFANTRY

In the British system, the standard infantry unit is the battalion of about 1,000 men, which is part of a larger organisation, the regiment. Regiments of infantry were never tactical organisations, but rather, administrative groups, each of which had the task of providing and maintaining one or more battalions. Until 1881, the only regiments of the British Army which fielded more than one battalion, in normal times, were the Foot Guards, the two Rifle Corps, and the first twenty-five regiments of the Line. Then the remaining single-battalion regiments were linked to form new regiments each of two battalions. The regiment continued to be a social grouping for purposes of mutual support and *ésprit de corps*. Its battalions did not normally serve together either in battle or in garrison, and in this respect the organisation of the British infantry differs fundamentally from that of most other armies, where the infantry regiment is a fighting unit, with its two battalions serving together as its component parts.

Battalions were commanded by lieutenant-colonels. The rank of colonel was, notionally, that of the commander of a regiment. There were two majors, each of whom could command a 'wing', or half-battalion, if it became necessary to divide the battalion, as it frequently was in India.

There were ten companies in a battalion until 1862 when the élite grenadier and light companies were disbanded and the total reduced to eight. The company, about 100 strong, was commanded by a captain, with two subalterns, each of whom was responsible for a half-company. The company was normally divided into four sections, each commanded by a sergeant, and the captain was assisted in his administration by a colour sergeant.

In 1912 this system was altered, and the companies were paired to form four large companies each of 200 men, as had long been the fashion in the major European armies. The four senior captains became majors commanding companies, and the four junior ones became second-in-command of companies, a newly created post, analogous to the battery captains of artillery. The four senior colour sergeants became company sergeant-majors, and the juniors became company quartermaster-sergeants, thus establishing a lower-level headquarters to take account of the increased administrative burden of the company commanders. The company was divided into four platoons of about 50 men each commanded by a lieutenant or second lieutenant, with a sergeant as second-in-command, and each platoon was further divided into four sections each under a corporal.

Between 1860 and 1922, there were normally fifty battalions of British infantry serving with the Indian Army at any one time. Battalions served tours of up to sixteen years in India, during which time they were kept up to strength by drafts from their linked battalions in the United Kingdom. As was the case with the cavalry and artillery, about one-third of the British Army's infantry was always stationed in India.

The East India Company's European infantry was organised in the same way as the British. By 1857 the army of each presidency included three single-battalion regiments. Four of these were designated 'fusiliers' and two were 'light infantry', but in fact these titles were essentially honorific; they all had the same battlefield role, that of 'heavy' infantry. In common with other units employed as shock troops, the Company's European infantry enjoyed a reputation for ferocity on the field of battle combined with brutality off it. During the Mutiny, three extra regiments were raised for the Bengal Army, but these were later disbanded, and the original nine battalions were added to the British Line and numbered the 101st to 109th Regiments of Foot.

INDIAN INFANTRY

The organisation of the Company's Indian infantry was based on that of the British. From 1824 onwards regiments consisted of one battalion only, divided into ten companies. Five companies were commanded by captains, and five by lieutenants. Each company commander had one European subaltern, and two Indian officers (a subedar and a jemadar) to assist him. By 1857 the Bengal Army had seventy-four regular native infantry regiments, the Madras Army fifty-two, and the Bombay Army twenty-nine. In addition to these, each army included several battalions of irregular infantry, in the frontier forces, or the contingents of local troops.

Irregular infantry battalions were each allotted three European officers. These were the commandant (normally a major or captain), the second-in-command (a captain or lieutenant) and the adjutant (a subaltern). As with the irregular cavalry, the system was popular with the officers, who held a more interesting and responsible post than if they had stayed with their parent regiments, together with the extra allowances this brought; with the soldiers, since it meant that Indian officers acted as company commanders and lieutenants, with consequent prestige and responsibility; and with the Government, which officered extra regiments at very little extra cost, since vacancies thus created in the establishments of the regular regiments were not filled up. The silladar system, however,

Royal Artillery: heavy gun in elephant draught. Photograph c 1880

Sepoys of the light company, 65th Bengal Native Infantry, 1846. Lithograph by J. Harris after Henry Martens

British officer, havildar and sepoys. 19th Bombay Native Infantry, c 1848. Note the regimental band (usually Eurasians) in the background. Lithograph by J. Harris after Henry Martens

was not adopted by the irregular infantry, except in the case of the two regiments of Jacob's Rifles. Ever logical, John Jacob adopted it for his frontier force riflemen, and even advocated the organisation of silladar artillery.

After the Mutiny, the infantry of the Indian Army was reorganised on a modified irregular system. Each regiment was to consist as before of one battalion, made up of eight companies as in the British Army. Each company was under a subedar, with a jemadar as his second-in-command, and was made up of 5 havildars (sergeants), 5 naiks (corporals), 2 drummers and 75 sepoys. There were 6 European officers at battalion HQ (the commandant, second-in-command, adjutant, quartermaster, medical officer and one general duties officer) with 2 officers as wing commanders, each in charge of a 'wing' consisting of four companies. The quartermaster was an ordinary regimental officer, not a former sergeant commissioned from the ranks as in the British service. Unlike the cavalry, Indian infantry continued to be clothed, armed, and maintained entirely by the State.

The number of British officers was twice that allowed to the old irregular corps, but the new establishment was found to be too small. In 1874 one or two 'probationers' or junior officers in their first years of appointment were added to each unit. Several reverses during the Second Afghan War were directly attributable to the casualties among officers inevitable in major operations. A regiment with two or three officers out of action had lost up to half its officer strength, and the Indian officers used as company commanders had not been trained to take over the full range of leadership duties. Extra officers were therefore sanctioned, to give each of the wing commanders a British subaltern.

In the 1890s, the Indian infantry replaced the unwieldy 'wings' or half-regiments by four 'double-companies'. These formed sub-units with a British major or captain in command, and another captain or subaltern as second-in-command, to which subalterns in training were attached, and were thus large enough to support a sufficient number of British officers to withstand normal casualties. As before, each company had its own Indian officers, a subedar and a jemadar, each of whom in battle commanded a half-company of two sections. The section was made up of twenty sepoys under a havildar. The establishment of an Indian infantry battalion at the beginning of World War I was, in most cases, 14 British officers, 16 Indian officers, and about 900 rank and file. This compared with a British battalion establishment, in India, of 28 officers and just over 1,000 rank and file.

The 1922 reforms brought the Indian infantry more closely into line with the British. The system of four large companies was adopted, each under a British commander and second-in-command, as in the old 'double-companies'. There were four platoons in each company, two under a subedar and two under a jemadar, each assisted by a havildar. Every platoon was divided into four sections, made up of a naik or lance-naik, and about ten sepoys. There was also a headquarters wing, made up of specialist sections, the band, signallers, machine-gunners, police, artificers, cooks and drivers.

The infantry in 1914 had been made up of 129 battalions including twelve pioneer units. The 1922 order of battle included 140 battalions, including 13 of pioneers, but with a reduced establishment of only about 750 men in each.

The personal armament of British infantrymen, until the middle of the nineteenth century, was the flintlock Tower musket, or 'Brown Bess'. This weapon was replaced by the Minié muzzle-loading rifle at the time of the Crimean War, and the Enfield rifle, a reduced-calibre variant of this, first introduced in 1853–4, was subsequently adopted for general service. The Indian Army was also issued with this weapon, in exchange for the old musket, to ensure that it kept up-to-date with the most modern equipment. It was the introduction of this weapon, which from the lubricants used in its cartridges threatened ritual pollution to Hindu and Muslim sepoys alike, that sparked off the Indian Mutiny. The bitter fighting of that campaign, in which some sepoys used their new rifles against their old masters to good effect, made the British determined that never again would the Indian soldiers be equipped with a weapon of the same fighting value as that of the British.

Accordingly, after the Mutiny, the Indian soldier was issued with an improved weapon only after the British had discarded it in exchange for an even better one. Thus the offending Enfield rifle was withdrawn in favour of the identical weapon without rifling, a smoothbore Enfield musket in fact, with less range and accuracy than the rifled version. The Snider breech-loading rifle began to be issued to British units in India in 1866, and the British troops' Enfield rifles were handed over to Indian troops. In 1874 the Martini-Henry rifle reached the British Army in quantity and the Sniders were transferred to the Indian infantry, and were retained by them until 1892. In that year large quantities of the bolt action Lee-Metford rifle were supplied to the British Army, and the Martini was given to the Indian troops. In 1905 the Martini-Henry was discarded in favour of the bolt-action Lee-Metfords and Long Lee-Enfields. The determination to issue to Indian troops an inferior weapon still persisted, however, and even as late as 1911, when the standard British infantry weapon was the Short Magazine Lee-Enfield, a special model was designed as a single-shot, non-magazine weapon for the Indian Army. This, however, was not issued, and during World War I the majority of Indian soldiers were at last equipped with the most modern rifle available.

Until the Indian Mutiny, the official policy was to assimilate the dress of the Indian troops as closely as possible to that of the British, for whom they were intended to be a substitute. Thus in 1857 the regular Indian infantryman appeared in a scarlet coatee, with white bars across the chest, white cross-belts, tall, black shako-like head-dress, and tightly cut trousers, white or dark blue according to the season of the year.

After the Mutiny, the Indian-style clothing of the irregular infantry regiments (many of which had been hastily formed to deal with the crisis) came into general use. As so often in the history of military fashion, the success of soldiers in the field leads to the success of their tailoring. Bernard Shaw, noting that after the Franco-Prussian War the British Army ceased to wear the shako (identified with the defeated French) and replaced it by the spiked helmet (the head-dress of the victorious Germans), supposed that the success of the Turks in any major war would be followed by the general adoption of the fez. So in India, the shako gave place to the turban, and the light trousers to loosely cut knickerbockers and puttees. The coatee had already been abandoned by the British Army in favour of the full-skirted tunic, in obedience to the dictates of contemporary military and civilian fashion. Indian infantry adopted either a tunic of European cut, or a slightly looser garment, made rather in the style of a shirt or blouse. Scarlet continued to be the predominant colour. Rifle regiments wore dark green. 'Khaki', or dust colour, was worn by the frontier force units and several others originally formed as irregulars.

Khaki was worn by all troops, British and Indian, between 1861 and 1864, for hot-weather duties. It was then replaced by the white cotton which had been worn previously, but it was re-issued for hot-weather operations during the Second Afghan War. Thereafter, white was worn for hot-weather parades and walking-out dress only, but the coloured serge uniforms remained in use in cold weather (except on operations) until 1914.

4

The Supporting Troops

The fighting soldier, in every army, must rely upon the services of other men, in less glamorous roles, to provide him with a 'life support system'. His most important requirements include the provision and supply of arms, equipment, clothing and food, and medical care for himself and his animals. These services were provided in India in a number of different ways, and were for the most part administered by the military secretaries of each government rather than the commanders-in-chief.

Ordnance

The ordnance services dealt primarily with the provision of arms and military stores. Each of the three presidencies evolved its own ordnance department at the end of the eighteenth century from earlier magazine establishments of the artillery. In fact from the very earliest times until 1922 the officers of the ordnance department were drawn exclusively from that arm, and no officers of other corps were admitted.

Various supply centres were established in each presidency, located in many separate areas, according to the needs of the troops at the time they were set up. However, changing political circumstances led to a situation in which by the second half of the nineteenth century the actual location, content, and duties of these arsenals, magazines and depots, as they were variously called, had little relevance to contemporary military requirements. There were no less than thirty-four such ordnance supply centres, with a further ten factories for the production of guns and carriages. Gunpowder and ammunition were manufactured locally to a certain extent, but most of the supplies of these, with the greater part of the small arms, were obtained from the United Kingdom.

The duplication of resources in each presidency was not entirely without good cause. Each army had evolved separately and needed its own stores under its own control. Communications, even between different parts of the same presidency, were bad. Before the coming of the railways, heavy stores could be moved no faster than bullock carts could pull them. And in times of drought or famine animals could travel only for so many days before they had consumed all the forage they could carry with them. In such circumstances their 'pay load' was actually nil, and, assuming a beast carried ten days' forage, this meant it could go only five days into a stricken area before starting the five days' march out again. Another important factor was security. The dispersion of depots, and in particular the separate resources of the three presidencies, meant that in the event of a sudden and unexpected disaster, such as the Mutiny of 1857, striking one area, the others would still be able to supply armies independently.

By 1875 the importance of both these factors had diminished. The fixed proportion of British to Indian troops (about one to three or four) then insisted upon, the concentration of artillery in British hands, the arming of Indian troops with weapons inferior to those of the British, all made any repetition of the serious position of 1857 unlikely to recur. The steady growth of the railway system enabled troops to be moved rapidly to a threatened area, and at the same time meant that stores from any ordnance depot in the sub-continent could be dispatched to troops requiring them, either for war or for their ordinary peace-time requirements, in a matter of days.

After a number of official inquiries and reports the whole of the ordnance services were reorganised in 1884 and amalgamated into one single Ordnance Department. A post of Director-General of Ordnance in India was created, with three Inspector-Generals of Ordnance under him each of whom controlled all the supply centres and factories in his own area. These coincided with the areas of the presidency armies which at this time were still separate bodies. The number of establishments was reduced to twenty-six, and nomenclature was regularised by the use of the terms 'arsenal' and 'depot' to mean large and small establishments respectively, with the restriction of the term 'magazine' to regimental or station arms and explosives stores. Proper equipment tables were prepared for units, and the distribution of reserve stocks between units and ordnance was laid down at agreed scales. The system of internal procedure and organisation was standardised, and definite principles were adopted for the calculation of reserves of armaments and equipment at depots.

In 1898 the policy of concentration was carried a step further. The small arsenals at Bellary and Mhow and the depots at Trimulgherry and Mandalay were closed, and arsenals at Bombay, Fort William (Calcutta), and Karachi were reduced to depots. The depots at Rawalpindi and Quetta, which had been established during the Second Afghan War, became the major supply arsenals of the North-West Frontier, and a new arsenal was built at Kirki.

This geographical division of the Ordnance Department was replaced in the years immediately before 1912 by a functional division. One directorate was established to deal with manufacture (the factories), another to deal with stores and distribution (the arsenals and depots), and a third was created to deal with inspection (which had previously been undertaken by the factories themselves).

The factories at this time included a gun and shell factory at Cossipore, near Calcutta; one small-arms factory at Dum-Dum, near by (hence the name of the 'dum-dum' bullet, originally a standard 0·303-inch bullet with an exposed lead tip, to give it increased 'stopping power' against fanatical tribesmen) and another at Kirki; a cordite factory at Wellington; harness and saddlery factories at Cawnpore and Madras; a gun-carriage factory at Jabalpur; and at Ishapur, in place of the old powder mills, a rifle factory and steelworks. An experimental and research establishment was set up on the coast at Balasore, where the long expanses of sand revealed at low tide made a suitable range for the conduct of ballistic tests and the recovery of projectiles. However, these factories had an output sufficient only to maintain peacetime stocks and to undertake minor campaigns. The Indian Army continued to be dependent on the United Kingdom for heavy equipment and for the supply of stores when they were needed in very large quantities.

In peacetime all heavy equipment, siege trains, engineer stores and ordnance parks were kept in store at the arsenals. On mobilisation, equipment was issued to units requiring it, and the field ordnance parks were sent out to join the troops on service. Supply was then from the field parks to units, and the parks in turn were resupplied from the depots and arsenals further back. These were manned and officered by artillery personnel, and formed part of the artillery component.

With the Kitchener reforms each arsenal was given one or two of the permanent divisions to supply. Their commanders or deputies were made members of the divisional staffs with the appointment of Assistant Directors of Ordnance Services, with a dual responsibility both to their own ordnance superiors, and to the generals commanding the divisions which they supported.

Food and clothing were the responsibility of the commissariat departments first established as such in 1810. The burden was made less onerous by the arrangements whereby silladar cavalry regiments were self-maintained in all respects except firearms and ammunition, which came from the ordnance.

Commissariat

In normal circumstances all Indian troops supplied their own rations under local unit arrangements, and so it was only in certain isolated stations or difficult country that they might be a responsibility of the official army commissaries. On active service, basic rations were issued, but wherever possible the Indian troops were expected to supplement them from local sources out of their own pay.

The British troops' diet was predominantly meat (usually beef) and bread, baked from wheat flour with yeast, in the conventional European style. Indian troops' diet varied in accordance with their personal tastes and prejudices. The staple ingredients were *ātta* (flour) and *ghī* (clarified butter). In place of bread they ate *capātīs* (unleavened pancakes). *Dāl* (pulses or lentils) was another standard ingredient, and so was rice. The diet, although mainly vegetarian, was not exclusively so, and meat was eaten by most troops. The Muslims were not allowed by their religion to eat pork or ham. The Hindus were forbidden by theirs to eat beef. Mutton was the usual meat dish of Indian troops. To avoid ritual pollution it was usual for each sepoy to prepare and cook his own food. This rule, however, could be relaxed and central cooking adopted, provided that the cooks were of the same class and religion as the sepoys concerned.

The commissariat's main concern was with the provision of items such as rations for European soldiers, grain and fodder for government animals (such as gun-teams and British cavalry horses, but not the animals of silladars or transport contractors), fuel for cooking and heating, bedding and accommodation stores, uniforms, personal equipment, and boots. The clothing of European troops was issued through unit quartermasters, and although the individual soldier was responsible for ensuring that it was not lost, misused, damaged, or neglected, it remained the property of the State. That of the Indian soldier was sold to him, repayment being made out of extra allowances given for this purpose, and became his personal property. In both cases damage by enemy action or fair wear and tear were made good from official funds.

Great reliance was placed on the principle of local purchase or the placing of special contracts, as the simplest and most economical method of obtaining supplies. This system, on the whole, worked well in the more civilised parts of the sub-continent, where there were usually adequate supplies of local materials and labour to work with it, and where the social and economic structure of the country included a vigorous class of entrepreneurs and merchant bankers. The profit motive could usually be relied upon to ensure that even in relatively backward areas, or on campaigns, contractors and civil labour would be available to provide what was wanted. However, such arrangements were not entirely stable, and a sudden deterioration in the military situation, particularly if communications were cut or if local goodwill ceased to be forthcoming, could result in inconvenience or hardship for the troops.

On campaign, commissariat officers were included with each formation, with the duty of purchasing and distributing the supplies which the army needed. They were assisted by a subordinate staff, also from the commissariat department. But there was no permanent machinery for the maintenance of troops in the field, and the necessary supply units had to be hastily formed on mobilisation. Thus the normal peacetime system was put under a strain, since the commissariat service had to continue supplying the troops not ordered on campaign while at the same time detaching personnel to staff the new field force units. Meanwhile, the columns actually on service suffered inconvenience while the newly formed supply teams settled down to the efficient performance of their duties, for even when officers and men had been on campaign before and knew what was required of them, most would now be filling different offices in a different set of conditions. This arrangement was one very typical of the Indian Army's support and administrative system. It was quite adequate for a peacetime colonial garrison, and even for small local campaigns. It was not designed to deal with prolonged hostilities, and proved unable to cope easily with them when they arose.

The arrangements that had evolved by the beginning of World War I included the establishment of supply officers at every important military station. The officer in charge of supplies was responsible to the station commander for the control and supervision of all supply matters. He arranged contracts and made purchases, and maintained a small supply depot. These depots were limited in their contents to items that could not be obtained either by local purchase or contract. Sealed patterns of items were deposited at the larger gaols (where convict labour was used to manufacture clothing supplies) and at major textile manufacturers, to prevent delays which had arisen in previous crises from the continued queries and references sent back to Army Headquarters about details of design. Mobilisation depots held separate stores of clothing and food for use when required.

Transport

Despite the importance in all campaigns, and especially those in India, of properly organised transport, no permanent transport department was established in the Indian Army until 1884. Prior to that time, reliance was placed on the time-honoured principles of using civilian contractors and resorting to local purchase. Unit transport was under regimental arrangements. Silladar regiments provided their own integral transport at no extra cost to the Government. Other units, in peacetime at least, needed little transport when in cantonments, and hired it as necessary from local businessmen who specialised in this service. When regiments went on long marches, or changed locations, the necessary carts and transport animals, with their drivers, were hired from local firms in the same way. Traditionally, Indian merchants travelled in caravans for mutual protection, especially in areas where robbers were active. The animals which carried their loads were hired with their drivers for such journeys, and, at the destination, fresh loads were taken on from merchants wishing to make the journey in the opposite direction. The Army, then, in adopting this system, was merely following the normal practice of the country. Trained animals and expert drivers were available, and it was very convenient and simple to make use of them when required. The Army was not obliged to feed or take care of either men or animals. The driver might be expected to look after his beast, as his livelihood depended upon it, and the contractor and caravan-master to look after both, as their incomes too depended upon satisfactory and continuous provision of their services. The Army was thus spared the necessity of maintaining in peacetime large numbers of men and animals whose services were required only in war, but who nevertheless required at all times to be fed, watered, exercised or paid, according to their several necessities.

As in so many other cases, the system, suitable for peacetime and local campaigns in settled areas, was less useful if subjected to unusual strains. There simply was not the spare capacity available to meet any sudden demand for expansion. The firms which held the ordinary peacetime contracts for army transport had no more interest than the Army itself in maintaining beasts and men not in permanent use. In fact, as profit-making concerns, they had even less interest in such an idea. Other transport firms could be brought in only at the expense of disrupting ordinary civilian commercial traffic. Moreover, as service with the Army in wartime was likely to be at best arduous, at worst actually dangerous, as well as less profitable, few such firms were eager to transfer their business in this way. The latter difficulty, however, was surmounted by the ancient right of Indian governments to press into military service transport animals and their owners, in return for the usual peacetime rates of hire.

Other measures were taken to fill the need. Wherever possible, animals were purchased and men enrolled as drivers to attend to them. Here, too, supplies were limited, the civilian firms themselves having resorted to this device before the Army realised that they alone could not cope with the demands. Sometimes animals were brought from considerable distances, but all this took time, not only to assemble them in the right place but also to train them after their arrival.

In some cases men and beasts were brought in from other occupations. This was the case during the Second Afghan War, when operations in the mountainous and desert regions across the frontier required large quantities of transport for the movement of supplies for the Army. Contractors offered considerable financial inducements to land-holders and men of substance throughout north-western India to obtain camels and bullocks normally used by the cultivators on their estates as draught animals in their ordinary work. The cultivators themselves were obliged to give up their animals, which were owned by the landowners. Then, being unable to plough or otherwise work the land, they avoided destitution by accepting employment with their beasts as drivers. Thus the transport lines came to include men who were not soldiers, and not even drivers, and who had no wish to be either, but who were forced to accompany the Army on its campaigns out of sheer economic necessity. They had no relish for the nomadic existence, no training, and no discipline to enable them to overcome the problems involved. Nor should the hardships suffered by the animals be forgotten. Like their men, they needed to be trained for war, and to operate as a team or part of a convoy. The ordinary farm animal would have been used to life as an individual, or at most with a few other beasts, and in many cases could have counted on regular feeds, water, and a comfortable byre. The lack of food and water on long marches was a new and uncomfortable experience to beasts as well as men. They felt the heat and cold as much, and feared the noise of battle perhaps with greater terror. Many suffered at the hands of drivers who were thoughtless, ignorant, or desperate. Literally thousands died, and the loss to the Army and the disruption to the civilian community was very great. This was one of the several factors leading to the establishment of a permanent transport department after the Second Afghan War.

The separate transport departments were amalgamated with the existing commissariat departments in 1887, and the three commissariat-transport corps were amalgamated into one corps for the whole of the Indian Army in 1889. The 'Rice Corps', as it became popularly known, thereafter took an important part in all Indian campaigns.

Between 1899 and 1904 a further reorganisation took place. The unorganised and improvised transport arrangements under regimental or depot control were replaced by regular corps and cadres of camel, mule, and cart transport under British officers. A further ninety officers were added to the corps, drawn from both the British and Indian services.

There were 21 mule transport corps, 18 mule transport cadres, 9 silladar camel cadres, and 2 pony-cart train cadres. Each mule corps had a British commander and was divided into two subdivisions, each under a warrant officer. The corps intended to support cavalry brigades were divided into six draught and four pack-troops, each under a dafadar (Indian sergeant), and had a total strength of 552 all ranks and 936 mules. Those supporting infantry brigades were divided into nine pack-troops, each under a dafadar, and had a total strength of 388 all ranks and 840 mules. The cadres also had a British commander, with an establishment of warrant officers, NCOs and tradesmen, but with a reduced number of mules and drivers.

Each silladar camel corps cadre had a British officer in command, and was made up of four subdivisions, each under an Indian officer. It had an establishment of 405 all ranks, and 357 camels. On mobilisation the number of camels was increased to 1,068. Four extra camel corps were raised by a revival of the *jagirdāri* system, in which Crown revenues were alienated in return for military service. These were located in the area of the Punjab irrigated by the Chenab canal. These districts had previously been desert and uninhabitable. When the canal was constructed it was possible to colonise them and the land revenues of certain districts were granted to camel-owners in return for their keeping a fixed number of animals available for military purposes when required.

Each corps, or cadre, was given a permanent headquarters, situated in most cases in the areas which supplied their additional animals on mobilisation. Enumeration officers were appointed to list the resources of animals in the country, and to keep registers of what would be available in an emergency.

Table 2 sets out some of the characteristics of the types of transport most commonly used. The speeds given in the table are those usually achieved on good and level roads. In hilly country, and across difficult areas, only half these rates of progress could be expected. In comparison, infantry marched at about three miles per hour, and cavalry, at the trot, seven miles per hour.

The pack-transport required special care and organisation. Loads had to be properly balanced and secured, to ensure that the animals did not develop sore backs and that the movement of columns was not disrupted by loads falling off and blocking the road. Whenever possible, animals were not loaded until shortly before the beginning of each march, and the loads were removed at long halts. Special loading parties had to be organised to assist the drivers in order for these procedures to be carried out effectively.

One driver normally led two or three animals. Elephants were driven by their own

Table 2

Characteristics of transport used by the Indian Army

Description	*Speed in miles per hour*	*Number per minute passing a given point in single file*
Pack-mules and ponies	5	25
Horse- or mule-drawn carts and wagons	2½	10
Bullock carts	1½	6
Camels	2	10
Pack-bullocks	2	16
Pack-donkeys	1½	12
Coolies	2	30

personal *mahūts*. On easy tracks camels were often tied together in strings of six or eight, led by one driver, with another bringing up the rear. On rough hill tracks, mules preferred to pick their own routes, and usually went best if driven in bunches rather than led tied together.

The amount of road space required by each type of transport, including intervals, was roughly as follows:

Pack-animals (or pair)	4 yards
Camel (or pair)	5 yards
2-pony or mule cart	7 yards
4-horsed wagon	15 yards
2-bullock cart	10 yards
4-bullock wagon	20 yards

Each gun and limber, accompanied by its ammunition wagon, occupied about forty yards. Cavalry, in sections of four, required four yards per section. Infantry, in fours, required two yards for every four men.

Medical

The Indian Medical Service was formed in 1896 from its three predecessors, the Bengal, Madras, and Bombay Medical Services. These had been organised in the 1760s to bring into a common organisation all the surgeons and physicians in the East India Company's employment. This included not only the military officers, but also those holding official posts under the civilian authorities. The Indian Medical Service, therefore, although primarily a military body, played an important part in the civilian medical administration. All its officers served for a few years with the army, and could then apply for transfer to civil employ.

The doctors who served with the army could be employed in either staff, specialist or regimental duties. Each regiment had an establishment of one medical officer. His essential task was to treat the wounded in time of war. The Indian climate, however, meant that he was often equally busy in peace and war, attending to the sick. For this reason, much attention was given to the problems of preventive medicine. The Indian Medical Service included in its ranks many men who made great contributions to the study of hygiene and tropical diseases. One such was Sir Ronald Ross, who discovered that men contracted malaria through the bite of the anopheles mosquito. Anaesthetics came into use in India in the late 1850s following their adoption in Europe, but the importance of antisepsis was not fully realised for at least another decade. Amputation was the most common treatment adopted in the case of severely wounded limbs, but when surgery was carried out in non-sterile conditions a large number of patients subsequently died from blood-poisoning. Doctors in India were as skilled as their colleagues in other countries, but the unwholesome climate, bad sanitation, and abundance of flies meant that their patients were at even greater risk than elsewhere.

Admission to the Indian Medical Service was restricted to qualified medical practitioners. Under the Company's rule appointments were part of the patronage of the Directors. Thereafter they were by open competitive examination.

The British regiments in India were cared for by their own medical officers, who were part of the British Army and did not belong to the Indian Medical Service. Although in emergencies members of the two separate corps could work together, the normal task of the IMS was to attend to Indian soldiers, their European officers, and other European officials in the Government service. The wives and families of civil and military officers were also treated by members of the IMS.

Subordinate medical departments existed, staffed by Indian, Anglo-Indian, and European practitioners, whose medical qualifications were not among those recognised in the United Kingdom. The details of their organisation changed from time to time. In 1822 a 'native doctor' was attached to each regiment, to treat those sepoys who did not trust the methods of the European doctors. In 1922 there were senior assistant surgeons (European and Anglo-Indian) who held the ranks of lieutenant, captain, and major, and who served in the station hospitals of British troops, and sub-assistant surgeons (Indians) who held the ranks of jemadar, subedar, and subedar-major and served in the station hospitals of Indian troops. Before World War I there were no hospitals for Indian troops other than their regimental medical centres, and the subordinate Indian practitioners were employed in British establishments as military hospital assistants. As was the case with the IMS, the subordinate medical departments provided staff for a number of civil posts, who thus furnished a reserve in time of war.

The Army Bearer Corps was formed in 1902 to bring the various regimental ambulance establishments under one central control. The method of collecting and evacuating wounded, however, remained much the same. Although stretcher-bearer parties were not unknown, the usual procedure was for a wounded man to be carried by a comrade, or make his own way, back to a 'doolie', a covered litter capable of carrying one or two casualties and carried by teams of bearers. These then made their way back to the dressing-station, where the regimental doctor was based. If a column could not leave its sick and wounded behind after a battle, they were carried with it, in doolies, in hospital wagons, or in litters on pack-animals.

Nursing the sick and wounded was for the most part a duty performed by military nursing orderlies. In the case of Indian units these were drawn from the combatant personnel of the regiment, and received no training other than that given them by the medical officer in charge. A similar system was in force in the case of British regiments until 1855. Then (as part of the reforms in army medical administration following the scandals revealed during the Crimean War) the Medical Staff Corps, later the Army Hospital Corps, was formed. This corps had no officers of its own, as the doctors were organised into a separate Army Medical Department, and it was not until 1898 that the two were combined into the Royal Army Medical Corps. The RAMC and its predecessors provided the male nurses and ward orderlies at British station hospitals. Regimental medical centres continued to be staffed by orderlies drawn from the soldiers of the unit which they served.

The first official use of trained female nurses in Indian military hospitals was in 1887. Lady Roberts, wife of the Commander-in-Chief, India, had for many years felt that the nursing care provided for British officers and soldiers in India was of too low a standard. She was, however, unable to make any alteration in the system until her husband reached a position of such eminence that, as a member of the Governor-General's Council, he could draw her plans to the attention of the highest authority in India. The then Governor-General, Lord Dufferin, accepted the proposal that a number of trained nursing sisters should be brought from the United Kingdom to work in Indian military hospitals. The Secretary of State agreed, subject to one proviso. Official funds would pay for the passage of the nurses to India, and their salaries while serving there. But they would not support the cost of frequent passages back to the United Kingdom on health grounds. The scheme would be accepted if homes for the nurses were established in hill stations where they might rest and recuperate during the hot weather, and if the cost of these homes was defrayed by public subscription. 'Lady Roberts' Fund' was subscribed to, generously, by every British regiment and battery in India, and within a few months the scheme was in operation.

A further need was for the care of young officers, convalescent after discharge from official hospitals. Many went to the hills to recover, but, lacking proper food and attention in the clubs and hotels where they stayed, subsequently died of their illnesses. A further appeal, to the officers, led to a fund which provided wards for such officers at the Lady Roberts' Nurses Homes and paid for nurses to attend them either there or in the plains when the hill stations were closed during winter.

About thirty members of the Army Nursing Service were sent to India at the end of the nineteenth century and carried on the work of the Lady Roberts' nurses. They formed the nucleus of the Queen Alexandra's Military Nursing Service for India, organised in 1903. The establishment of 'QAs' at this time was eighty-four nursing sisters, four lady superintendents, and a chief lady superintendent, who was the administrative head of the service. Their duties were to nurse sick and wounded British soldiers in hospital (including the British officers of Indian units) and to instruct and supervise the ward staff of the RAMC. They also controlled the hospitals where the wives and families of British troops were treated.

No female nurses were provided for Indian troops. In Indian society, women of respectable families lived a secluded existence, and the physical contact involved in nursing duties was not acceptable to the conventions of the time. Even among the most progressive individuals, there were few who wished their daughters to receive a western-style education. Until well after the period covered by this book, there were few Indian girls with the modern education required in a nurse's training, and fewer still whose families would have countenanced the idea of their leaving home to become Army nurses.

Curiously, the Afghans, who are generally regarded as among the most zealous in the seclusion of their womenfolk, were ahead of the Indians in the provision of nurses. The intelligence report on the Afghan forces at the battle of Ahmad Khel, 19 April 1880, contains the information that 'There was a very large number of real *Ghazis* [religious warriors] among the gathering, including twelve women who were admitted as *Ghazis* and were allowed to remove their *pardahs* [veils] on the condition that they followed the *Ghazis* into action, and took water, etc., etc., to the wounded.'

In both Europe and Asia a common belief in polite society was that to admit young women into military nursing would be a threat at least to their modesty, at worst to their chastity. The English, who gave their nurses the title 'sister', were concerned in the same way as the Afghans to suggest that their nurses' reputation was protected by some religious connection.

Veterinary

The Indian Veterinary Departments were first formed in 1826. Each cavalry regiment and artillery formation had its own veterinary officer who attended its animals when required. Like the medical officers, they accompanied their regiments on active service. One veterinary surgeon, serving with the 3rd Bombay Light Cavalry, actually joined them in a charge, during the Persian War of 1857, and captured an enemy standard. With the extension of the silladar system to nearly all the Indian cavalry, and the concentration of artillery into British hands, the Army Veterinary Service in India from the 1860s onwards was concerned almost entirely with the care of animals with British units. The silladar cavalryman was responsible for the care of his own horse without any assistance from the State. Transport animals were the responsibility of the civilian contractors who provided them, or of the army transport units, when these were formed. In 1914 there were a total of sixty-four veterinary officers serving with the Army in India, including a number from the British Army's Veterinary Corps, who served in India for a few years as an ordinary tour of duty.

Hospitals were established for the treatment of government animals. The first to be set up with an army in the field was in the Ethiopian campaign of 1868. During World War I the service expanded, and in addition to the field sections (or mobile veterinary teams) and base depot veterinary stores previously maintained, many other types of veterinary establishments were formed. Among these were stationary and mobile hospitals, special hospitals for camels, and convalescent depots, some for horses, others for camels.

Remounts

The Remount Department had its origin in the stud departments established in each presidency at the end of the nineteenth century. At this time most 'country-bred' Indian horses were small, and of poor physique, and this was an attempt to improve the breed and enable the Company to supply horses for its troops from its own dominions. The traditional source of supply of good horses to India had been Central Asia, but this was a source that was expensive and also one that could not be controlled in an emergency. Disruption of traffic through the Afghan passes, or indeed anywhere in territory not under British control, could cut off from their armies an essential military import.

After several changes in organisation the Remount Department was established to deal with the supply and breeding of horses for the Army. It supervised the breeding of horses, mules, and donkeys in some provinces, although in others this was dealt with by the Civil Veterinary Department. Considerable improvements were made in the strength and stamina of country-bred horses in the last quarter of the nineteenth century, and, as far as the requirements of Indian cavalry were concerned, the aim of diminishing the dependence of the Indian Army on imported animals was achieved. Young stock was purchased from breeders and reared on government farms, from which they were sold either to Indian silladar units or on the open market.

The British cavalry and artillery were supplied entirely by the Remount Department from official sources. The import of some Arab horses was continued, but a much larger number of horses was imported from Australia. These were called 'Walers', after the colony of New South Wales. Mountain artillery and non-silladar cavalry were also issued with animals by the Remount Department.

The establishment of animals which the Remount Department had to maintain amounted, in 1913, to the following:

British cavalry	5,049 horses
Royal Horse Artillery and Royal Field Artillery	14,845 horses
Mountain Artillery	3,760 mules
Non-silladar Indian cavalry	1,536 horses
Others	386 horses
	2,294 mules

A small war reserve of 500 cavalry horses, 500 artillery horses, and 200 ordnance mules was kept up, but there were no arrangements for supplying large-scale reinforcements in the event of a major expansion on mobilisation or to replace heavy casualties.

5

The Reserves

European reserves

BRITISH TROOPS

The arrangements for the provision of a reserve for the British component of the Indian Army can be described quite simply. They did not exist. The voluntary reserve forces of the British Army, consisting of the Militia, the Yeomanry, and the Volunteer Force, made up of patriotic citizens who trained in their spare time, were for home defence only, and could not be embodied except in the case of a national emergency. The Territorial Force, formed in 1908 from the Volunteers and Yeomanry, was subject to the same restrictions, although in World War I its regiments volunteered for overseas duty, and several were sent to India to replace regular units needed elsewhere.

Among these was the 25th London Regiment. This was a cyclist unit which had trained for many years to act as mobile troops. On the outbreak of war, in the usual humorous way in which the British Army treats its Volunteer soldiers, they were ordered to hand in their machines and to act as ordinary infantry. They ended their active service in Afghanistan, perhaps one of the countries in all the world least suited for the bicycle.

The Regular Army possessed no reserve of its own until 1870. The time-honoured practice of enlisting men for life, or until their services were no longer required, was replaced in 1847 by enlistment for periods of ten or twelve years at a time. In 1870 this was replaced by the 'short service' system, whereby men enlisted for a period of six years' full-time service, 'with the Colours' as it was called, followed by six years on the reserve, with liability to be recalled to active duty in an emergency, in return for a small retaining fee. Men were allowed to extend their full-time service only in special circumstances. On mobilisation, reservists reported to their regimental depots and were used to bring up to strength the regular units in the United Kingdom. These were normally considerably under-strength through providing drafts for service with units overseas, in India or elsewhere.

There were a few British military reservists actually living in India. These were usually time-expired non-commissioned officers or soldiers of good character who had elected to remain in India on the expiry of their term of service with the colours. It was official policy to encourage this, to build up the European reserve in India, and jobs were found for them on the railways and in the postal and telegraph departments. In this way the Government of India ensured that these vital communications were in reliable hands in the event of invasion or internal disorder. But their numbers were too few to make a significant addition to the order of battle, and on mobilisation they were required as lines of communication and depot personnel. Although they provided well-trained and skilled manpower which gave invaluable support to the army in the field, they still could not add materially to the number of fighting troops.

Even with the advent of railways and steamships, it still took at least eight weeks before any significant reinforcements could reach India from the United Kingdom, and for that length of time the British troops in India had to be self-supporting.

Therefore, contrary to all normal military practice, the British units in India were actually larger in peacetime than they were in wartime. The peace establishment had to contain a sufficient margin to ensure that on mobilisation, after a unit had left behind its sick or medically unfit, plus detachments for other duties, it was still strong enough to appear in the field at full war strength. Each unit would leave, in a safe garrison in India, men unfit for active service, the wives and families who had accompanied the troops to India, a small clerical and stores section to act as a link between the unit in the field and its home depot, and sufficient men to share in their local defence in case of disorders. Men sent back from the front or rejoining from hospital were held there, and when fit for duty, were sent on to the fighting unit together with new drafts.

RESERVE OF OFFICERS

The lack of reserves of British soldiers was matched by a lack of reserve officers. Retired British Army officers did not settle in India, nor, generally speaking, did British officers of the Indian forces. Units of the British Army going to India, whether for routine tours or after mobilisation, had their own officers, regulars or reservists respectively. The Indian Army Reserve of Officers, in 1914, had forty members.

At the beginning of World War I, indeed, the Indian Army had to supply officers for the British reserves. In August 1914 about 10 per cent of the officers of the Indian Army were on leave in the United Kingdom. They were at once seconded to the British Army to help train and command the new armies raised by the Secretary of State for War, Lord Kitchener, a former C-in-C, India. These new armies were made up of units formed especially for the duration of the war, from the hundreds of thousands of recruits flocking to the colours. But the brave performance of these new armies in battle does not mean that equal or better results could not have been achieved without departing from the existing Territorial system for expansion. And it may well be that the 550 Indian Army officers concerned would have been better employed with their own regiments, where they were sorely missed, than with Kitchener's new units.

The Indian Medical Service was deliberately intended to have both civil and military functions in order to secure a reserve of doctors. All its members, after qualifying as doctors, were commissioned as officers in the Army, and had to carry out a tour of some years' regimental duty before becoming eligible for civil employment, with its greater attractions. Civil appointments included posts in charge of the health and hygiene of districts or stations, local medical administration, and superintending or teaching at government hospitals. Although not all were available for immediate military service on mobilisation, a high proportion of them were, and in wartime, when the medical branches need to expand more than any other service, the number of doctors available for service with Indian regiments could be virtually doubled, although at the expense of the civilian organisations.

The engineer officer reserves consisted of two sections, those still serving with the Army, but not with formed units, and those in the civil departments of the Indian Government. The first section dealt with military works services, ie, all those tasks connected with the construction and upkeep of the Army's garrisons and cantonments, their roads, bridges, buildings, drains and water-points. These officers generally served alternate tours with the sappers and miners and with the works services. All, therefore, were instantly available on mobilisation, to bring the sappers and miners up to war strength. The second part consisted of officers seconded to civil departments such as the Survey of India, Railways, Public Works Department (mainly on roads and canals) and Mints.

The normal Indian tour of a Royal Engineer officer was for about five years, but if he found the conditions of service there attractive, and was himself considered suitable, he might continue to serve there on a permanent basis. A major attraction was that by doing so he not only received a higher rate of pay (especially if serving in a civil appointment) but qualified at an earlier age for a retirement pension paid at Indian (ie, expatriate, and therefore, enhanced) scales. In 1880 there were about 70 engineer officers serving with the Army in India, and 180 with the civil departments. In 1910 the proportion had been reversed and was 130 to 70, but there was still a significant reserve of officers there.

This system, like that of the medical service, meant that civil departments were paying for the Army's reserve, men who might disappear in time of emergency, when they would be most needed by the civil administration to deal with the inevitable emergencies of wartime. While every employer is inconvenienced by the loss of reservists, he is not normally expected to employ serving soldiers, and this system, although useful to the Army, was unpopular with the civil administration, and became increasingly so with the growing interest of Indian nationalist politicians in the subject.

THE INDIAN VOLUNTEER FORCE

The only formed bodies of European troops available to support the regulars were provided by the Indian Volunteer Force. This corresponded to the Volunteer Force of the United Kingdom. It was raised from public-spirited citizens of European or mixed descent who trained in their own time and mostly at their own expense.

In general, fashions among the European community in India followed those in England with a few years' time-lag, and this tended to apply as much to the fashion for volunteering as any other. But the wave of volunteering which swept the United Kingdom during the invasion scare of 1860 was actually preceded by a few years in India, when men hastened to form Volunteer Corps after the outbreak of the Indian Mutiny. In the towns and cities caught up in the Mutiny, the Company's civilian employees once again had to take up arms for their own defence and to support the hard-pressed British and loyal Indian forces. The garrison at Lucknow, for instance, consisted of 133 British officers, including many whose regiments had mutinied, 671 British regular soldiers, 51 Eurasians, mostly drummers from mutinous Indian regiments, and 153 Volunteers, with 712 Indian soldiers. The Volunteers, hastily raised on the outbreak of disturbances, consisted mainly of civil servants, with a sprinkling of merchants and others who chanced to be in Oudh at the time and who made their way to Lucknow for safety. Some of them formed a troop of volunteer cavalry, which proved more than a match for the regular light cavalry of the mutineers. The rest were drilled as infantry by sergeants of the 32nd Foot, or served as volunteer artillerymen manning the fixed defences. Although the infantry were the despair of their instructors when it came to manoeuvring as a formed body, some being bent with age, others corpulent, others mere youths, and some considering it beneath their dignity to be drilled with muskets like common soldiers, they proved capable of putting up a stiff resistance behind a parapet. Some had heavy sporting rifles of their own, which enabled

(above) *A sepoy, in 'regimental Mufti', going on furlough with his wife and child. 34th Sikh Pioneers, c 1913. Watercolour by G. F. Paterson;* (left) *A sepoy foraging. 34th Sikh Pioneers, c 1913. Watercolour by G. F. Paterson*

Gurkha officer and riflemen. 6th Gurkha Rifles, 1904. Note the rifleman's kukri, or Gurkha knife, on his right hip. Watercolour by A. C. Lovett

Sepoy on guard. 124th Duchess of Connaught's Own Baluchistan Infantry, 1904. Note the gong, struck to signal the end of each tour of 'sentry-go'. Watercolour by A. C. Lovett

them to make good practice against enemy sharpshooters. Others had shotguns which had a devastating effect at close quarters against enemy storming parties.

There was at first some criticism that the Volunteers followed the example of the regular soldiers in the garrison, men of the 32nd Foot, in their excessive and intemperate consumption of alcohol. However, this gradually died out, either because supplies of liquor ran down, or as men realised the seriousness of this breach of the military code at such a time. One of the keenest Volunteers in the garrison of Anderson's Post, at the southern corner of the defences, was Signor Barsotelli, a dealer in alabaster, from Florence. Armed with a musket, a double-barrelled rifle, and a huge sabre, he first went on duty with his cartridge pouch slung round his neck like an Italian organ grinder. When it was pointed out that the conventional way was to wear the pouch in the middle of the back until required, he expressed considerable pleasure, saying it was difficult enough to avoid tripping over the sword. He soon adopted the wiles of a hardened old soldier. Working on an age-old military principle, he told a fellow Volunteer, who was doubtful of his ability to present arms on the nocturnal approach of the orderly officer on his rounds, 'Never mind, sir, make a leetle noise. Who will see in the dark?' Nevertheless he proved a valiant and noble soldier, manning his post with great courage during the thickest of the fighting.

In the great presidency cities and provincial capitals, other Volunteer Corps were formed. They were intended to defend the localities in which their members lived, and although they did not join the armies in the field, they did free the sorely needed European regular troops for this purpose. It is doubtful if the Army ever left the defence of an area that was actually threatened to the Volunteers alone. Indeed, their presence was regarded by professional soldiers as at best a superfluity, and at worst a possible embarrassment. But the demands of European communities for troops to protect them, even hundreds of miles from the nearest mutineer, would have meant that valuable trained troops would have been wasted in this way, for political rather than military reasons. The Volunteer Corps were a means by which the Europeans in safe areas could have both the duty and the satisfaction of guarding their own homes and families, while the regulars were freed from the performance of an unnecessary task.

After the Mutiny, these emergency units were gradually stood down, only the Madras Guards, as they were later called, retaining a continuous existence until disbanded in 1947. However, in the 1860s fresh units were raised, often made up of men who had served in the earlier corps. Many included young men who had been Volunteers in the United Kingdom, and who went to India on commercial business as the European community there increased in size.

Most of these units were, like the Volunteers in the United Kingdom in the 1860s, recruited from the middle classes, or from men at least of sufficient affluence to invest time and money in their uniforms and equipment, since at that time no payment was made out of official sources except after actual mobilisation. In the United Kingdom this gradually changed, so that ordinary working men could join and actually receive pay or expenses incurred in training, and accordingly only a few units retained a socially exclusive nature. In India there was virtually no European working class; the European community was mostly bourgeois or middle-class. (One aristocratic Vicereine, Lady Curzon, speaking of the necessity she had to mix with Indian official society, remarked that she was determined not to be fastidious.) Indeed, like the regular military officers, most Europeans in India, whether in government or commercial employment, enjoyed a far higher standard of living there than that to which they could have aspired in England.

Each corps had a regular adjutant and sergeant-major provided by units of the British Army serving in India, and all Volunteers came under command of the general officer commanding the military area in which they were based.

The Eurasian community found its military place in the units sponsored by the major railway companies. As we have seen, many time-expired British soldiers took service with the Indian Railway. Often they married Indian wives, and a high proportion of jobs in the railway service was reserved for men of Anglo-Indian descent. Partly this was a recognition by the British Government of its social responsibility towards these people, who did not fit in with the structure of Indian society and tried to identify themselves with the Europeans, who rather looked down on them. Partly it ensured that in any emergency or nationalist rising these vital communications were in the hands of men loyal to the British connection. The formation of railway battalions to guard the depot installations and permanent way against terrorists, and to defend the cantonments and civil lines against rioting mobs (for this was, with the growth of the Indian nationalist movement from the 1880s, a more likely threat than attack by enemy armies), gave this community a chance for military service. The right to bear arms, under British rule, was severely restricted, and gave considerable status to those to whom it was granted. These railway units of the Volunteer Force gave the Anglo-Indians some satisfaction at being given the chance to defend their homes and livelihoods, against 'the natives', who were not allowed to join.

Joining their respective unit was in fact made a condition of employment by the railway companies on their British professional engineers and superintendents. This was not always popular, and the appointment of unwilling officers damaged the efficiency of some corps. However, in general the numbers and enthusiasm of the railway volunteers made them a valuable body for local defence purposes, and their familiarity with the layout of the installations they were defending, and with railway operating procedures, meant that interference with the proper running of trains was kept to a minimum.

In 1907 the Indian Volunteer Force amounted to 34,000 men, organised in sixty-one different corps. There were seven regiments of light horse, five garrison artillery units, and four battalions of mounted rifles.

During World War I the Volunteers were embodied from time to time and took over garrison duties normally performed by regulars. Coastal units especially were mobilised in the early part of the war, when German cruisers were still at large in the Indian Ocean, and Indian shores were liable to be bombarded. However, as the force was restricted to local service, and had not become a general-purpose force like the British Territorials, its units could be deployed not where the C-in-C India wanted, but only where its members had agreed to serve. Its major contribution to the war effort was probably in supplying men to the Indian Army Reserve of Officers. The Volunteer who knew Hindustani, could show familiarity with military practices, and generally had some aptitude for leadership, was ideal material to refill the decimated ranks of Indian Army officers. The patriotic spirit of the time was such that all who could be spared, and many who could not, joined the armed forces, and a high percentage of the 1,000 men who joined the Indian Army Reserve of Officers and saw active service during that war were former Volunteers.

Public opinion in India demanded a reform of the system, and the 1917 Indian Defence Force Act introduced compulsory military service for European British subjects in India. Unlike the conscription system by then in force in the United Kingdom, men were not required to serve overseas, nor were they called up for continuous training, but nevertheless in an emergency they could have been embodied for service anywhere in India. All the Volunteer corps were incorporated into this Defence Force, in which Indian citizens for the first time were allowed to serve as Volunteers.

As part of the postwar reforms an Act was passed in 1920 establishing a body known as the Auxiliary Force (India). This absorbed the European and Eurasian units of the Indian Defence Force. It consisted of ten cavalry units, variously called light horse or mounted rifles, five brigades and three independent batteries of field artillery, four companies of Royal Engineers and thirty-four battalions of infantry including ten railway regiments. Service in this force was to be of a strictly local or provincial character.

As before, these units came under the local military authorities for operations. Regular officers were appointed as adjutants to each regiment, battalion or artillery brigade. Training was carried on throughout the year, as with the British Territorials, but took place in the early morning rather than the evening. Pay was given for each day's training, and a bounty was paid to 'efficient' Volunteers on completion of a fortnight's annual training in camp.

Indian reserves

REGULARS

A regular reserve system was not introduced for Indian troops until 1886, a few years after it was adopted by the British Army. Originally it provided for two groups, an active reserve and a garrison reserve. The active reserve was composed of men with a minimum period of five (later reduced to three) years' colour service. Unlike the British Army's scheme, this did not make reserve liability a compulsory part of the contract of enlistment. Indeed, reserve service was not open to men who had completed more than twelve years with the colours or reached the age of thirty-two (the usual age of recruits was about eighteen). Active reservists reported to their regimental centres for one month's training each year. The garrison reserve, consisting of men pensioned after a total of twenty-one years' regular or reserve and regular combined service, was discontinued after a few years, as the plan to call them up for one month every other year was not enforced, and the advanced age (in contemporary Indian terms) of the men meant that there was no useful role which they could perform on mobilisation.

Kitchener's reforms of 1904 tried to make the reserve more efficient. Its establishment was to be gradually increased from 25,000 to 50,000, although a cautious Finance Department insisted, in return, that the reserve pay was cut from 3 rupees per month to 2. Silladar cavalry was brought into the scheme. Arms and equipment were kept in arsenals, and personal kit at the regimental centres, for issue on mobilisation or for training purposes. Unlike the British reservist, who had no liability for further training, his Indian comrade was required to perform one month's training annually, later to be altered to a two-month period once every two years. This was easier in Indian conditions, since after leaving the

Army most men returned to their own villages, as peasant proprietors, so that they were free to absent themselves once each year's harvest was in, for a considerable length of time. After a total of twenty-five years' service, regular or combined, reservists were granted a pension of Rs 3 or 3·8 (3 rupees 8 annas) per month without further liability. Service continued to be voluntary, and although men from the Punjab and north India willingly accepted it, it proved less popular in the Bombay presidency, and was never strongly taken up in Madras.

The scheme, nevertheless, was not entirely satisfactory. It consisted only of one class, so that the Government could not call out selected groups but only the whole at once. There was not the machinery to allow a flexible use of manpower, calling up those with recent service before those who had long quitted the colours, for instance, or younger men before older men.

A further important factor which would have affected any reserve scheme is that, in India, men once retired from active employment seem to age much more rapidly than is common in the West. The fit man of thirty was commonly, ten years after quitting the colours, a dotard of forty (an age which is still a good average life-span for inhabitants of an undeveloped country). The decline was accelerated by the social environment of the soldier, drawn, as most were, from agricultural villages. Able to live in modest comfort in the large family system, the reservist could sit in the shade and watch his grandchildren play around him. In the worst cases senility occurred at an incredibly early age. In others, even if their martial spirit was unquenched, physical debility had set in. One aged veteran, called up for service in 1914, arrived at Marseilles in December as a putative reinforcement for the Indian Corps in France. Asked by the inspecting medical officer about his teeth, he replied that, by the mercy of God, he had still one upper tooth left. He then proceeded to remove it and present it to the startled doctor for inspection. But drafts of such men were a useless encumbrance to the Army in the field, and it proved a waste of precious time, money, and supplies to attempt to use them.

LINKED BATTALIONS

Prior to 1886 there was no machinery by which one unit could be reinforced by men of another, except by persuading individuals to volunteer for transfer. However, in that year, following the example of the British infantry, Indian infantry units were linked to form groups of about three battalions. Enlistment was for service in any one of the battalions in the group required by the Government, but transfers between units would be enforced only if one or more of them were actually ordered on active service. In 1888 regimental centres were allotted to each group, and it was planned that one battalion should always be there, to give the men a tour of duty periodically near their homes, and to be available to find drafts for the others if required.

The disadvantage was that units could be reinforced only at the expense of others in the same group. In 1914, it was found that forty-three fully trained battalions, or about one-third of the total strength of the Indian infantry, would have had to be reduced to cadres if the needs of those proceeding on active service were to be supplied from those stationed in the group centres. The system could work only for limited warfare, involving only a small proportion of the Army at any one time. There was no group depot in the sense of a common source from which all the units in the group were administered, but only a common geographical location, at which each separate regiment established its own separate depot.

In 1922 this was remedied by the introduction of a true regimental system in the Indian infantry and pioneers. The new regiments consisted of about four regular battalions, plus a training battalion and a Territorial Force battalion. The previous system had allowed each unit to keep its separate title and distinctions. These were now abolished. Thus, to take an example, the units previously known as the 30th Punjabis, 31st Punjabis, 33rd Punjabis, 9th Bhopal Infantry, and 46th Punjabis became respectively the 1st, 2nd, 3rd, 4th, and 10th Battalions of the 16th Punjab Regiment. The 10th Battalion of each regiment was permanently allotted the role of a training battalion, stationed at the regimental depots. However, something of the old system remained in that these training battalions were composed of companies on the scale of one for each of the active battalions in the regiment. Each active battalion provided the officers, instructors, and administrative staff of its own company of the training battalion, and this company was also organised, in peacetime, to undertake the accounting, records, and other base activities of its affiliated battalion in war, assisted by any clerical and headquarters personnel of that battalion not proceeding on war service. However, recruiting and training were controlled and organised by the training battalion commander on behalf of all the units in the regiment. Recruits were enlisted for service in any unit of the regiment at any time, and although in peacetime each battalion drew recruits from its own training company, in war reinforcements could be sent from any section of the training battalion as a whole. Alternate tours of duty with the training battalion and with the line battalions did much to create a common regimental spirit among officers and men, making for ease of cross-posting in time of need.

The ten regiments of Gurkha Rifles had, since 1903 (and in some cases before that) been composed of two battalions in the same way as the regiments of the British Army. This arrangement was left undisturbed, and each depot continued to train recruits for its own regiment as before.

Reserve service was not made compulsory on Indian soldiers of the other arms. This was because they were, in the main, organised as mobile units. In the circumstances of the time, this meant they had large numbers of horses or mules on their establishments. In the United Kingdom it was possible to requisition animals from civilian sources on mobilisation. In India, with certain exceptions, this was politically impossible. Therefore the Army had to maintain in peace sufficient animals to enable it to mobilise, plus a reserve to replace immediate casualties. The money to pay for the improved infantry reserve had been found by reducing the number of men actually serving with each battalion in peacetime. This could not be done with the other arms, as the number of men required to look after the horses could not be reduced, and so there was no money available for increased expenditure on reserves.

The cavalry regiments, which had previously been quite independent, were in 1922 linked in groups of three. Their numbers were reduced by amalgamations from thirty-nine to twenty-one, and thus there were seven groups. Seven stations were selected as suitable locations for regiments allotted to internal security duties, and at each of these stations a group had one of its three regiments. These were not made training units, and no peacetime depots were created. But in war a group depot would be formed out of the surplus personnel of all three regiments and reservists of the group as a whole. One of the defects of the previous system was that regiments had tended to recruit from special communities favoured by their own officers. Reinforcements in war could be provided only by drafts of suitable men and their horses (which they owned under the silladar system) from other units, and it was virtually the case that the mobilisation of one Indian cavalry regiment meant the demobilisation of two others. The 1922 reforms required all regiments in a group to recruit from the same classes, to allow for mutual support.

THE INDIAN TERRITORIAL FORCE

There was no opportunity for public-spirited Indian civilians to train as soldiers in their spare time until the Indian Territorial Force was formed in 1921. This was one result of the Montagu-Chelmsford reforms of the Indian constitution which recognised self-government as the goal of British policy in India. Indian politicians claimed that self-government involved self-defence, a view which the British Government itself had adopted in withdrawing its garrisons from white colonies as soon as self-government was achieved. This transferred the burden of defence from the British to the colonial taxpayer, who thus had no representation without taxation. British reluctance to hand over control of Indian defence was seen, quite correctly, as reluctance to hand over control of government.

The Indian Territorial Force, therefore, was intended partly to be a sop to nationalist opinion. It allowed the chance of part-time military training to the growing Western-educated urban classes who were the very last people whom the British wanted in the regular army. It was organised into two categories, Provincial battalions and University Training Corps battalions. The Provincial battalions were trained and organised as units of line infantry, and were allotted to the new infantry regiments, when these were formed, numbered as their 11th or 12th battalions. There were no arrangements for local training throughout the year. Instead each battalion was embodied once a year for four weeks' training at the regimental depot.

Training was provided by regular instructors or pensioners specially employed for the twenty-eight-day periods. The senior officers were British instructors from the depot battalion. The junior officers were Indians, but to begin with they held only honorary commissions as captains and lieutenants in the Indian Land Forces. Their powers of command derived only from 'Viceroy's commissions' as Indian officers of the Indian Territorial Force, and they had no powers of command over European troops.

The University Training Corps battalions were modelled on the Territorial contingents of the Officers Training Corps at British universities. The liability for actual war service was not intended to be enforced, and the units were meant to act as a link between the military and academic worlds.

It was hoped that the volatile character of the Indian student population might benefit from this opportunity to acquire the military virtues of discipline and responsible leadership. Graduates would have the opportunity to serve either as officers, or (since the number of commissioned appointments was so small) soldiers, in the Provincial battalions, which were open to any citizen who cared to join. Corps were raised at the universities of Bombay, Calcutta, Allahabad, Lahore, Madras, Rangoon, Patna, and Delhi.

However, the old objection, that any money spent on volunteer units could be better spent on regulars, continued to be raised. In January 1923 the then C-in-C India, Rawlinson, was arguing that troops in India had to be available as much for duties in aid of the civil power as for defence against a foreign invader. The Indian Territorial Force could not, in practice, be used to quell riots, any more than the Territorial Army (as the former Territorial Force had been renamed) in the United Kingdom. If cuts had to be made in the defence estimates, he said, the Territorials rather than the regulars should take them.

THE INDIAN STATES FORCES AND IMPERIAL SERVICE TROOPS

The Indian princes, ruling their own states under British influence, were allowed to keep up small armies of their own. They were guided in their policies by British Residents or Agents at their courts, who discouraged unnecessary military expenditure. But it was necessary for a ruler's social position to have at least a large bodyguard, and the employment of men in military service enabled a ruler to give a gainful occupation to many of his loyal subjects, which he was expected to do in return for their support.

During the Second Afghan War, some of the native states in the Punjab sent their own troops to support the regulars, and these did useful service in the Kurram Valley campaign. Subsequently, in 1885, when war with Russia over Afghanistan seemed imminent, many princes, in a display of loyalty, offered troops to the Government of India. From these offers the Government formed, in 1889, the Imperial Service Troops. Princes wishing to join the scheme undertook to raise and maintain units trained and equipped on the model of the Indian Army. Each ruler paid for his own troops, who were recruited from his own subjects, were commanded by his own, Indian, officers, and belonged entirely to his own establishment. A British officer was detached from the Indian Army to act as 'Military Adviser' to each ruler, with other officers as Assistant Military Advisers if the numbers of troops justified their presence.

The establishment at the beginning of World War I was about 22,000, made up of two mountain batteries (from Kashmir), four companies of sappers, fifteen cavalry units, three camel corps, thirteen infantry battalions, and seven transport corps.

In the postwar reorganisation these troops were renamed Indian States Forces. The emphasis was now less on the defence of the Empire of India and more on the role of the states in any future federation when British India became self-governing. The establishment was increased as the princes saw that it would be to their advantage to have well-trained troops, and attempts were made to bring unit strengths and armaments into line with those of the Indian Army. Some regiments were maintained as militia, but most remained full-time regular regiments, available either for service with the Indian Army or to deal with any internal disorders that might occur within their own state. The total strength at the end of 1922 was about 7,500 cavalry, 15,500 infantry, 700 artillerymen, 800 sappers, 750 camel corps troops, and 1,750 transport personnel.

Commissioned service in these units allowed scope for Indian gentlemen who were denied the chance to serve in the Indian Army other than in the ranks. The State Forces were thus a valuable safety-valve, absorbing the military energies of able men who might otherwise have used their energies against the British regime which excluded them from honourable office. In the same way Indian statesmen and administrators could find an outlet for their talents in the native states, which was a major reason why the British, after 1857, left them alone. But there were few facilities for the instruction of officers in their professional and technical duties. The high incidence of palace guard duties and ceremonial functions on the birth, marriage, or death of members of the ruling family disrupted such training programmes as there were, and the limited promotion prospects implied by the small size of most state forces gave little incentive to officers to improve their standards, and they were frequently criticised for slovenliness, unmilitary attitudes, and general incompetence.

To sum up, the organisation of its reserves was in line with the rest of the Indian Army's administrative arrangements: inadequate for prolonged major operations, but quite suitable for its primary role as an imperial garrison force, for frontier wars and internal security emergencies.

6

The Indian Soldier

Indian manpower was an essential element in the British conquest of India, and its subsequent garrison. Without it the British Indian Empire could not have been gained, and without it the British position in India would have been untenable. In 1857 British control was threatened when the Bengal Army mutinied, and was largely retrieved by new armies raised in the Punjab. Constitutional progress towards self-government was accelerated by the part played by Indian soldiers in the two world wars, fought, as it was said, to defend the right of the self-determination of peoples. Indian soldiers, like Indian civilians, came to expect the right to control their own destiny, and the British could no longer rely on them to act indefinitely as an army of occupation in their own country. Indian soldiers, like the soldiers of other lands, generally shared the cultural and social background of their fellow-countrymen, and what was important in Indian society as a whole was also important to the army from which it was drawn.

The caste system

An essential feature of Hindu society was the division of its members into four great classes (*varṇa*), fundamental, pre-ordained, and divinely sanctioned. These classes, those of the *Brāhmaṇ* (priest), *Kṣatriya* (lord), *Vaiśya* (merchant) and *Śudra* (serf), were mutually exclusive. A man could not rise from the class he was born in, nor could a high-class man marry, take food from, or otherwise socially mix with, one of a lower class without being ritually polluted. Those outside the Hindu system were *Mleçcha* or untouchables, who performed menial or degrading tasks such as those of sweeper, or leather-worker (the cow being a holy animal to Hindus). Christians and Muslims, no matter how exalted they might be in rank and power, were also outcaste, ritually unclean, and so avoided by respectable Hindus as far as possible outside official duties.

Alongside this class structure grew up a complicated system of castes (*jāti*), the term given to the multitude of exclusive groups, based partly on region, partly on race, partly on occupation, into which Hindu society divided itself. Unlike the four classes, which were developed by the early Āryans and became part of the Hindu religion, the castes grew up later, as a way of enabling primitive or numerically weak peoples to adjust to successive waves of immigration and settlement, the introduction of new crafts, and the evolution of more complex economic or racial systems. Castes were governed by local committees of their older and respected members. They had the power to make rules for the caste, and judge those who infringed them. Serious breaches of caste rules, such as association with members of a caste thought to be lower in the social scale, might lead to expulsion which was one of the greatest calamities that could befall a Hindu. A man who was ritually polluted might not only lose his place in the cosmic order, his class, but also his place in society, his caste, and could thereafter mix only with the lowest of the low. From this it can be understood how important his caste was to the orthodox Hindu and how the soldier, who was but an ordinary man under his uniform, had to consider this in the performance of his duties.

Castes were not immutable, and their rules and customs might be varied from time to time. New castes might form and old ones disappear. Their relative place in the social hierarchy might change. But the caste system, with its own organisation, outside the power of the State, was one of the most powerful elements in the survival of Hindu culture. Under foreign dominion, whether Muslim or Christian, Mughal or British, the Hindu clung to, and was cared for by, his caste. The loyalty that he could no longer give to his country, or kingdom, or city, when that was the agency of a foreign ruler, was given to his caste, and thus reinforced its position as a powerful element in society. The advantages of the caste system, especially to poorer folk (to whom it gave a sort of social welfare scheme, helping the needy, caring for widows and orphans, providing a sure place in society, a fixed employment or means of livelihood), were so attractive in India that other, non-Hindu, inhabitants tended to adopt it. The Muslims, despite the fierce insistence of the Prophet on the equality before God of all believers, formed groups with many of the characteristics of a caste. The Sikhs, whose religion was the result of attempts to rid Hindus of caste prejudice and synthesise their faith with the monotheism of Islam, formed castes, despite the opposition of their religious teachers. The Christians, too, formed sects resembling castes. Converts to Roman Catholicism in the Portuguese settlements kept their prejudices, and high-caste converts kept apart from low-caste converts. The British ruling class formed a caste of its own, although never formalising it as such. The British members of the Indian Civil Service who ruled the country were themselves referred to, only half in jest, as 'white Brahmans'. At the other end of the Indian social scale, a caste was formed in southern India called the 'Quinsaps', consisting of families whose men served in the Queen's Own Madras Sappers and Miners, and who gave their daughters in marriage only to one another.

Although it was to the *kṣatriya* class that the sacred laws assigned the role of a warrior, the Hindu armies of India were from the earliest times made up of men from many different castes. There is evidence that there were, in the armies of various early rulers, permanent divisions and regiments, with their own corporate existence, standards, and property. But in no case does it seem that there was a caste or even a class barrier against men serving as soldiers. As we have seen, many groups of invaders were slotted into the class system as *kṣatriyas*, even though they were originally Āryans neither by race nor by culture. The Rajputs of western India, for instance, whose name in later times was almost synonymous with that of *kṣatriya*, were descendants of later invaders. Brahmans were allowed to serve as soldiers, or in other professions, as not all members of that class could find employment as priests.

The British in India raised their early armies, in the same way as other rulers, from Muslims or Hindus of any grade in society or any ethnic group who cared to take military service under them. In the main, because in any society men tend to follow an occupation which is familiar to them or their relatives, the soldiers who joined the British armies were from the same families that had served Indian rulers. In some cases, as when the British took over a province or kingdom, they took over its soldiers, either as individuals (when a man might find his employer had gone out of business, so to speak, and would join a rival firm), or wholesale (when the British raised new regiments to give employment to out-of-work soldiers who would otherwise have posed a risk to the security of their rule or the property of their subjects).

These soldiers, however, seem to have joined the army at most out of family tradition, rather than as a hereditary caste occupation. The idea that there were special 'martial castes', whose prime occupation was soldiering, and who were loyal, like the cat, to the house rather than its master, was a later concept, and the early British Indian armies were essentially recruited on an individual rather than a class system.

Commanding officers in all three armies, Bengal, Madras, and Bombay, tended to prefer men from the agricultural areas of the northern plains, especially Oudh and Rohilkhand, the area called Uttar Pradesh. These men were taller than most inhabitants of southern and eastern India, and since at all times it was the ambition of their British officers to have their sepoys appear as imposing as British troops, physical stature was a desirable attribute. Also it may have been thought that in hand-to-hand combat a tall man had the advantage over a short one. Cornwallis spoke highly of such sepoys. Writing to the Duke of York in 1790 he said how impressed he was with 'the native black troops—fine men and would not . . . disgrace even the Prussian ranks'.

Unfortunately these men also came from one of the most caste-conscious areas of India, and in the Bengal Army at least were able to bring their caste prejudices with them.

The extent to which the caste system was a vital part of Indian society meant that it could be a potential source of trouble. The Army, in effect, had to walk a tightrope. On the one hand it could fall into the trap of conceding every claim made on caste or class grounds, to the detriment of discipline and military efficiency. On the other hand, it disregarded at its peril the religious laws and customs of the community of which the soldier was a member, and which acted on him with as much force as on his fellow-countrymen.

In the Bengal Army during the first half of the nineteenth century, concession to caste feeling became excessive. The multiplicity of castes sanctified by custom and endorsed by religion allowed the soldier to use men of low classes for special tasks that they alone were permitted to do. Such restrictive practices resulted in soldiers on guard duty employing a special man to strike the gong, used to signal the time to change sentries, and another to carry the sentry's equipment for him. High-class sepoys declined to take orders from officers of lower class, with the result not that the former were punished, but that the latter, in Bengal at any rate, ceased to be promoted. Soldiers declined to dig trenches and earthworks at sieges, on the grounds that to do so was to perform the work of coolies and thus lose their own caste.

Yet, given leadership and mutual respect, such a situation was not bound to occur. The Madras Army, although the Vellore garrison had mutinied in 1806 against what they considered attacks on their caste, continued to recruit, and promote, men of low caste, or even of no caste at all. The Bombay Army even recruited men of the same families as the Bengal Army, but made no allowance for caste prejudices, and high-caste and low-caste men served together in the ranks, the low-caste men taking on the esteem due to fellow servants of the State, and fellow members of the regiment.

Something of the latter spirit was evidenced by Maharaja Sir Pratap Singh, regent of the state of Jodhpur in Rajasthan. In 1907 a young British officer at his court died. Only three other Christians, also British officials, could be found to bear the coffin, so the old regent himself made up the bearer party. The next day a body of Brahman priests waited on him. They began their humble address with the observation that yesterday a dreadful thing had happened. The Maharaja agreed; he had lost a true friend. The priests went on to venture the opinion that still worse had occurred, that the Maharaja had, by touching the coffin of a dead man (and a *Mleçcha*, or one outside the Āryan caste system, at that), become ritually polluted and had lost his own caste. The old warrior, with some force, expressed the view that as far as he was concerned the highest caste of all was one to which he and the dead Englishman both belonged, that of the soldier. The priests felt it prudent to withdraw, in haste, and the matter was not raised again.

During World War I, a Gurkha regiment's subedar-major authorised the issue of corned beef to his Hindu soldiers under the title corned mutton, trusting that the gods of their Hindu pantheon would forgive the offence committed by men whose only alternative was starvation.

A lost patrol captured by the Turks in Palestine was made up of Sikhs and Muslims. The Sikhs would have been slaughtered out of hand, but for the testimony of the Muslim corporal at their head that, on his oath sworn on the Koran, these Sikhs albeit uncouth and hairy, and from a distant part of his country, were as good Muslims as himself, and so should be treated as prisoners of war. The Sikhs themselves reported this, when a British attack enabled them to escape the next day.

Jemadar Sher Khan, a Muslim officer of the 125th Napiers Rifles, collected from Peshawar early in 1918 the widow and two little sons of his friend and comrade Subedar-Major Munshi Singh, a Hindu officer of the same regiment, who had been killed at the battle of Sannaiyat in 1916. He brought them down to Bombay to attend a ceremony at which the subedar-major's posthumous Indian Order of Merit was awarded. The acceptance by a Hindu widow of the protection of a Muslim man on a long and difficult journey was a sufficiently rare occurrence to be remarked upon as evidence of the mutual trust and respect which men of different religions could form for each other in the same regiment. Jemadar Sher Khan, too, was to fall in battle in Palestine later that year. Both officers were as popular and well-respected with each other's co-religionists as with their own.

Yet in 1857 it was over questions of caste that the Bengal Army finally mutinied. The army had been discontented for years. Although politically inactive, even the least patriotic sepoy must have felt that the British were going too far by annexing state after state in the Indian sub-continent. There was even a suggestion that the poor performance of some sepoy regiments in the Sikh Wars was the result of their reluctance to see the extinction of the last powerful independent Indian state.

The annexation of Oudh, where many sepoys were recruited, deprived them of the special privileges at the King of Oudh's court, accorded to British-protected persons. But in addition to the unease at the spread of British power, there was unease at the spread of Christian influence, either through the laws of British rulers based on Christian ethics, or through the direct efforts of missionaries. These had increased in numbers following the Charter Act of 1834 which withdrew from the Company the right to keep British subjects out of its dominions. Official chaplains had been provided for the Company's European servants from early times, but missionaries had been discouraged. In the eighteen-forties and fifties, however, missionaries of all sects proliferated.

It cannot be denied that in the fields of education, health, and social welfare, much missionary work was of benefit to the communities, but in too many cases it was combined with an intolerance and bigotry that generated more heat than light. Christianity in England was in its most confident and certain period. Hinduism in India was in a decadent and unreformed stage. Such a combination allowed the zealous missionary, even the zealous ordinary worshipper, to denounce the abomination of heathendom with fervour and energy. Major-General Wheeler (who was massacred with all his garrison and family after the surrender of Cawnpore to the mutineers) was only one of several commanding officers who had the disturbing habit of preaching to his sepoy regiment on parade. When it is appreciated that to the ordinary Indian the term Christian meant one who ate pork and beef indiscriminately, consumed alcohol intemperately, and was none too fussy about personal cleanliness, it will be realised with what horror the prospect of conversion to that faith was received in the ranks. The habit of the European soldiers of celebrating the nativity of their Redeemer by drinking to excess, even to insensibility, was a poor advertisement for the faith to which they were known to belong.

Other acts of the British were considered to threaten Hindu customs and hence religion. The abolition of *Thāgi* (from the practitioners of which the English word 'Thug' is taken) was denounced as irreligious, because these stranglers acted in the name of the goddess Kāli. The abolition of *Satī*, the burning alive of widows on their husbands' death, was also thought to be a blow against religion, although it had been condemned by the ancient Hindu law-givers. General Sir Charles Napier, the conqueror of Sind, agreed to allow the old custom of widow-burning out of respect for Indian feeling, but pointed out that it would be followed by the new custom of hanging those involved, out of respect for British feeling. Each side, he said, would thus respect the customs and feelings of the other. This, however, was not thought by many Indians to be a satisfactory way of looking at things. An Act permitting the remarriage of widows only made matters worse.

The sepoys were as concerned as civilians about such matters, that affected them all, but were directly affected by purely military reforms. One such was the General Service Enlistment Act of 1856. Tired of protracted wrangles over whether sepoys had the obligation to serve outside India, and of being obliged to call for volunteers and pay bounties to those willing to go, or of using force to put down mutineers among those who would not go when required, Government decreed that in future no recruit would be accepted unless he undertook liability for overseas service. This would involve high-class men in crossing 'the black water', or ocean, which was forbidden to Āryans on pain of loss of caste, or at least of expensive purification ceremonies. Rather than face these, young men would not enter the profession of their ancestors, and so lost an honourable and lucrative source of employment. Those already serving feared that they would be gradually pressured into accepting general service, and forced on board ships, where, if becalmed, they would be forced to eat salt pork and salt beef.

The pig, to Muslims, is an unclean creature. The cow, to Hindus, is a holy one. To either sect, therefore, the touch of a substance containing the fat of both animals is ritual pollution. To eat it, or put it to the lips, is attended by the greatest repugnance imaginable.

Yet such a substance was used, by the military authorities, to lubricate the paper cartridges used for the new Enfield muzzle-loading rifle first issued to Indian troops in 1856. It was widely believed that the nature of this compound was first drawn to general attention by a high-caste sepoy. He had scornfully refused to let a low-caste person drink from his water bottle, and had been told by the object of his scorn that soon no sepoy would have any caste at all, for the British had issued the new cartridges to pollute them all and turn them into Christians. The story spread like wildfire. The Indian soldiers came from a peasant class, and, like all peasants, preferred to believe what their friends and neighbours told them rather than what was announced by the authorities or even the evidence of their own senses.

In Bombay and Madras, where the weapons had not been widely issued, action was taken in time to stop alarm. Sir Patrick Grant, Commander-in-Chief, Madras, explained to his Governor, on 23 February 1856, that the drill for the new rifle required the soldier to bite off one end of the cartridge. The powder inside was then poured into the barrel, and the cartridge paper rammed down to form a wad in front of the bullet. The paper was greased for lubrication.

> The grease used for cartridges is tallow or lard; and there is no doubt whatever that applying the lips to such a substance is opposed to the religious prejudices of natives, Hindoo or Mahommedan.
>
> No order ought ever to be issued which cannot be legitimately and justly enforced; I consider that we are bound by every principle of policy and justice to abstain from any measure which is really opposed to the religious prejudices of the native troops; and I strongly recommend that immediate instructions should be issued prohibiting the issue to native corps of these greased cartridges.
>
> It has been suggested that *ghee* which every native eats might be substituted for tallow or lard; but I am sure this would not suffice and that it would be impossible to satisfy the minds of the men that *ghee* only and not grease of any other description was made use of.

The Madras Army was thereupon issued with ungreased cartridges, which men greased for themselves with *ghī* (clarified butter) or vegetable oils bought in the bazaar.

The Indian Mutiny

In Bengal the damage was done. Regiment after regiment refused the new cartridge. In vain did the authorities claim that the mixture used was of beeswax and coconut oil, in vain did they change the drill book and allow the cartridge to be torn by hand, in vain did they resort to the Madras expedient of authorising men to concoct their own lubricant. After several isolated cases where men had been discharged or degraded or imprisoned for refusing orders to use the cartridge, and where officers had been attacked, the garrison at Meerut (Mirath) rose in mutiny. Europeans, officers and civilians, with their families, were murdered, the prisoners released, and official buildings set on fire.

Many atrocities were attributed to the sepoys. Some tried to save their own officers, and to defend them from mutineers of other regiments, but some attacked and murdered them. Some protected the officers' wives and children and helped them escape, although some killed Europeans regardless of age or sex. Most men probably had an agonising choice between staying with their comrades or going with their foreign officers who had lost control. The worst atrocities seem to have been committed not by the mutineers but by local criminals and the city mob. Throughout northern India the scene was re-enacted, until at last the stream of British reinforcements poured in from the rest of India, from China, Ceylon, Singapore, and Europe itself, turning the tide, and British power was restored.

Atrocities and cruelties on the Indian side were now matched by equal ferocity on the

British. The mutineers, knowing that they had committed the ultimate military crime and that they could expect no mercy, had no choice but to fight on with their comrades until the end. Regiments kept together for the most part, under their own Indian officers, using in battle, to good effect, the same tactics and drills learned in British employment.

The British, frightened by the extent of the outbreak, and angered by stories of atrocities by the mutineers (at a time when Englishwomen received perhaps more outward respect and protection than at any other time in history) exacted a pitiless revenge. Most of the stories were afterwards found to have little foundation in fact, but in war truth is always among the first casualties. Vindictiveness pursued the mutineers even beyond the grave. Muslims were hanged with pig flesh crammed into their mouths. Hindus were forced to lick bloodstains from the site of murders and were then killed without the chance of purification. Mutineers were blown from cannon (an old Mughal custom) and their bodies so fragmented as to make decent burial impossible and even identification uncertain.

Although, amid this sad and tragic tale, there were acts of chivalry and kindness by fighting men and non-combatants of both sides, and although more Indian soldiers remained loyal than mutinied, 1857 marked a watershed in the history of the Indian Army. Never again were relations between Briton and Indian as completely cordial and unreserved as they once had been.

The Peel Commission, which was set up after the Mutiny to consider and report on the future organisation of the Indian Army, recommended that the 'Native Army should be composed of different nationalities and castes, and as a general rule mixed promiscuously through each regiment'. However, it did not specify whether it meant all nationalities and castes or only some. The 'mixed regiment' system of Bombay and Madras, whereby men of various religions and races served alongside each other in the ranks, was extended to the Bengal Army and Punjab Frontier Force so that their infantry, in 1864, consisted of twenty mixed regiments, sixteen regiments divided into companies each of separate classes, and seven class regiments (Gurkhas and Sikhs). In the Bengal Army, the suggestion that Christians and men of low caste should be recruited fell on deaf ears. In Madras and Bombay, however, low-caste soldiers, even untouchables, continued to be recruited.

After twenty years the Eden Commission, set up in 1879 to review the Indian Army, reported that the principle of divide and rule was being entirely disregarded. It urged that the regulations requiring Bombay and Madras regiments to recruit only from inhabitants of their own presidencies be rigorously enforced, and the practice of taking men from the Punjab (which had replaced Uttar Pradesh as the fashionable source of military manpower) should be stopped. If the two smaller armies could not recruit enough men of a type that suited them, they would have to be reduced in numbers until they could.

However, following the Anglo-Russian crisis of 1885, when it seemed for a time that sepoy and cossack would cross swords on the banks of the Oxus, the attempts to keep the Indian army regionalised were abandoned, and the whole emphasis changed to recruitment on a caste or racial system. This was because for the first time for two hundred years it seemed likely that the Indian Army would be called on to meet a European enemy, and it was judged that only certain races and communities of the Indian population were capable of meeting this challenge.

The racial criterion was a fairly new one. Previously men of a certain region had been sought after because of their height and build, or men had been excluded from service because they belonged to too low a social caste. Indeed, the caste-consciousness of the Bengal Army continued after the Mutiny. The Central India Horse, for instance, which, although a local corps, tried to associate itself with the more fashionable Bengal Army, discharged all its low-caste men as soon as possible after the Mutiny, even though those men, and that regiment, had been raised to deal with a crisis partly brought about by excessive indulgence of caste feeling.

The man under whom the new policy was brought into effect was General Lord Roberts, VC, Commander-in-Chief, India, from 1885 to 1893. He had been originally an officer of the Bengal Artillery, and although this corps had been amalgamated with the Royal Regiment after the Mutiny, and Roberts was by now technically an officer of the British Army, he was still prejudiced in favour of the Bengal troops. His reputation had been made during the Second Afghan War by his march to the relief of Kandahar, where the survivors of the disaster at Maiwand were besieged throughout August 1880. Roberts' account of the depressed state of the garrison was coloured, almost certainly unconsciously, to make the achievement of his own force stand out more boldly. The fact that the troops defeated at Maiwand and shut up in Kandahar belonged to the Bombay Army led Roberts and many Bengal officers (still smarting under the humiliation of their army by loyal Bombay troops during the Mutiny) to use this as proof that the Bombay soldiers were no longer of real military value. Indeed, Roberts wrote in his memoirs *Forty-One Years in India* (a book that became a best-seller in its day and was widely studied for information on Indian military questions) that before he started for Kandahar from Kabul he knew nothing of the condition of other troops available for its relief from Baluchistan 'except that, as belonging to the Bombay Army, they could not be composed of the best fighting races'. In a similar way the poor performance of some Madras Army units during the Third Burma War of 1887–9 was held to demonstrate that the Madrassis were of little use as fighting men, and that they should no longer be retained in the Army.

Roberts was Commander-in-Chief of the Madras Army from 1881 to 1885, but a closer connection with that Army's troops failed to change his views: 'I tried hard to discover in them those fighting qualities which had distinguished their forefathers during the wars of the last and the beginning of the present century. But . . . I was forced to the conclusion that the ancient military spirit had died in them.' He believed that he was being objective in this assessment, and went on to say: 'It was with extreme reluctance that I formed this opinion with regard to the successors of the old Coast Army, for which I had always entertained a great admiration. For the sake of the British officers belonging to the Madras Army too I was very loath to be convinced of its inferiority, for many of them were devoted to their regiments and were justly proud of their traditions.'

What Roberts might have considered was that it was these very officers who were the weakness, and not the material they had to command. Indeed, of the four Bombay regiments involved at Maiwand, three were recruited from the Punjab and the Delhi area, and it was the Grenadiers and the Bombay Sappers and Miners, both recruited from the Bombay presidency, who behaved most creditably.

After the Mutiny it had been decided deliberately not to employ regiments of one

The commanding officer, with his bugler, orderly and led charger, walking at the head of his men

The doolie or litter with occupant

The artist's own baggage initialled E.P.L. is carried by the first camel

Officers' mess stores

19th Bengal Infantry on the line of march, 1843. Lithograph after E. P. Layard

The 75th Foot at the siege of Delhi – Indian Mutiny, 1857. Campaign dress – shirt sleeves and forage caps. Lithograph by E. Walker after G. F. Atkinson

presidency in the territories of another except in cases of emergency. This was popular with the soldiers, who did not like tours of duty in what was to them strange and unfamiliar country, far from their homes and families. It was desirable from a political point of view, since it meant that if disaffection occurred among the troops of one region, there were troops in another who would have nothing in common with them, and so be readily available to put down disorders. The disadvantage was to restrict the chance of active service, except in rare major campaigns, to the Bengal Army. No ambitious young officer would join the Madras or Bombay Armies, where the country was settled and pacified, if he could go to the Bengal Army, with the northern frontiers to garrison, and the opportunity which every officer desires both to exercise his profession and to distinguish himself in it. Sir Bartle Frere, a distinguished Indian civil servant who rose to be Governor of Bombay, was one of several authorities who pointed out the age-old military truth that there are no bad soldiers, only bad officers, but this point of view was not held by Roberts and those who thought like him.

Sir Frederick Haines, a British Army officer with considerable experience of India, was well aware of the weakness of his officers when Commander-in-Chief, Madras, between 1871 and 1875. In September 1871, he wrote to the Commander-in-Chief, India, that this army had been

> so shut out, of late, from all participation of Field Service and Expeditions, that the officers and men are in danger of falling into the delusion that they have nothing but police duties to perform . . . This view of the functions of an army must tend to lower military spirit in all its members. I am therefore anxious to associate some of our officers from every branch of the service with the . . . camp of exercise at Delhi, when they will have an opportunity of studying their profession on a scale this Presidency has long ceased to afford.

From Madras he went on to become Commander-in-Chief, India (1876–81). In letters of February and March 1881 to the Duke of Cambridge, Commander-in-Chief of the British Army, he wrote that, so far from the population of Madras being essentially peaceful, there were many elements among it that could require military coercion in an emergency. He cited the Polygars of the south, Mysore (then about to be handed back to its native rulers), Coorg (where men habitually carried arms), Hyderabad (where the Nizam kept up a large mercenary army), and the Moplahs of Malabar (who even fifty years later were able to rise in armed and serious rebellion).

He was anxious to keep separate armies, each recruited from its own region with its own depots and factories for warlike stores, providing separate sources of military manpower and administration for the British to call on in case of internal disorders.

> It has been customary to declare that the Madras Army is composed of men physically inferior to those of the Bengal Army, and if stature alone be taken into consideration that is true. It is also said that by the force of circumstances the martial feeling and characteristics necessary to the real soldier are no longer to be found in its ranks. I feel bound to regret the above assertions and others which ascribe comparative inefficiency to the Madras troops. It is true that in recent years they have seen but little service, for, with the exception of the sappers, they have been especially excluded from all participation in work in the field. I cannot admit for one moment

that anything has occurred to disclose the fact that the Madras sepoy is inferior as a fighting man. The facts of history warrant us in assuming the contrary. In drill, training, and discipline the Madras sepoy is inferior to none, while in point of health, as exhibited by returns, he compares favourably with his neighbours. This has been manifested by the sappers and their followers in the Khyber, and the sappers are of the same race as the sepoys. I have no doubt all I have said for Madras may be urged equally justly for Bombay. I therefore have no hesitation in saying that the Madras and Bombay Armies form the safest and most efficient reserve for the army of Upper India.

Roberts thought that Madrassis, as a rule, were more intelligent and better educated than the races of northern India whom he considered born fighting men. He therefore proposed to take advantage of this intelligence to teach them the value of musketry and make them good shots. By implication he failed, but by implication too he failed because their officers lacked appreciation of the changes in fire control and tactics brought about as the musket was replaced by the rifle. 'It required time and patience,' he wrote in his *Forty-One Years in India*, 'to inspire officers with a belief in the wonderful shooting power of the Martini-Henry rifle, and it was even more difficult to make them realise that the better the weapon the greater the necessity for its being intelligently used.' The sappers and miners, he admitted, were indeed a most useful, efficient body of men (they were officered by the Royal Engineers, the corps which took the cream of the cadets from the Royal Military Academy), but he claimed that they were a brilliant exception to his theory, without going on to consider why this was the case.

Roberts argued that Asiatics, after long years of peace, became soft and unwarlike. Having been brought up in a prosperous and secure environment, the classes that had once been 'martial' had deteriorated, in his opinion, so that they could no longer be pitted successfully in war against other, still undegenerate, races of India, still less against Europeans. He argued that the introduction of the railway and telegraph, the high proportion of British troops in the garrison of India, and their control of all the artillery and fortresses, made the question of internal security increasingly irrelevant in the constitution of the Indian Army. Fighting value alone must be the standard to be applied. No comparison, he said, could be made 'between the martial value of a regiment recruited amongst the Ghurkhas of Nepal or the warlike races of northern India, and of one recruited from the effeminate peoples of the south'. He spoke of the 'erroneous belief that one Native was as good as another for purposes of war', and his despair 'at not being able to get people to see the matter with my eyes'. Nevertheless, supported by the Military Member, Major-General Sir George Chesney, he persevered, and had the satisfaction during his tour of command of mustering out several regiments and companies composed of men recruited from classes which did not meet his approval, and replacing them by those that did.

The theory of racial deterioration became widespread in British circles, and indeed in Indian ones. Other explanations, ranging from the semi-mystical to the quasi-scientific, were advanced to improve on Roberts' explanation of political environment. Some saw the enervating (to Europeans) climate of southern Indian and the Ganges delta as responsible for the character of their inhabitants, whom Sir O'Moore Creagh (Commander-in-Chief, India, 1909–14) described in his book *Indian Studies* as 'timid both by religion and habit, servile to their superiors but tyrannical to their inferiors, and quite unwarlike'. By comparison temperate-zone man, especially the European and Japanese, had cold winters and was of a warlike disposition. This philosophy, justifying both the rule of the British and their recruitment from selected classes, was a simple and popular one. Furthermore, it had a respectable antiquity. Political commentators since the days of Herodotus have agreed that hard countries breed hard men.

Another explanation for thc superiority of the 'martial' over the 'non-martial' classes was purely racialist: the Āryans had subjected the original inhabitants of the country and allowed only certain races to bear arms, so that in the course of time those races were the only ones fit to bear arms. A notable exponent of this theory was Sir George MacMunn, a Royal Artillery officer with considerable Indian Army experience, who as a major wrote *The Armies of India* (published 1911) and as a lieutenant-general wrote *The Martial Races of India* (published 1933). This explanation glossed over the historical fact that many of the 'martial classes' identified as such by MacMunn were not Āryans, or even Hindus, and that many communities that were demonstrably closest in culture to the classical Āryans were considered by him to be non-martial. The military achievements of the present-day Israelis, descended from the despised 'non-military' Jewish communities of Europe's cities, or the small, dark East Bengalis (who in the struggle to establish independent Bangladesh successfully defied the tall, fair-skinned soldiers of Pakistan, drawn from the 'martial classes' of the north-west) reveal the weakness of this argument. So did the distinguished service of Madrassis during World War II, especially as gunners, sappers, and signallers.

Nevertheless it was one flattering both to the British and the classes whom they selected as 'martial'. Communities vied with each other to be considered of the right stamp. Some, in their anxiety to ensure continued job opportunities for their young men, refused to serve with others of lower social caste, and put pressure on their British paymasters to discontinue their recruitment. Thus ironically the caste prejudices which were one factor in the decline of the old Bengal Army crept into the Bombay and Madras armies, as these were gradually cut down in numbers in accordance with the martial caste theory. The Maratha of Bombay came to consider that the outcaste Parwaris, and communities low in the social scale such as the Bhandari, the Koli, the Beni-Israel, and the Native Christian, were no longer fit to associate with him in the ranks. Even castes such as the Dhangar and Mali, who were clearly of Maratha stock, though not of the *Kṣatriya* class, were excluded from service, as much by the pressures of the Maratha soldiers as by their British officers. The Telegus and Tamils of Madras also had their service terminated, along with many other communities. They resented their exclusion as an insult, and feared it as a restriction on their chances of employment and respectable association with the ruling authorities. Indeed, the general attitude of excluded classes was not that they objected to the theory of martial castes as such, but only to the fact that they were not considered by the British to be among them.

As the proportion of Bombay and Madras regiments in the Indian Army diminished, so did the number of officers familiar with them and able to defend them against further cuts. MacMunn himself agreed that even the Bombay regiments which remained had ceased to recruit the 'mountain rats' that had constituted the conquering armies of the great Sivaji (who overcame the Mughals in the seventeenth century), preferring instead the taller men of the north in pursuance of the cult of grenadierdom. He conceded that one of the bravest soldiers he ever knew was Naik Anthony, an Indian Christian of the 23rd Wallajabad Light Infantry, killed in a sortie from a post on the Sino–Burmese border. He records how in 1920 the keen adjutant of a South Indian regiment implored him to use his influence to preserve the use of suitable manpower available from that area. Nevertheless MacMunn shared the prevailing military opinion that better soldiers could be found elsewhere for the same amount of tax-payers' money. Basing his experience on the poor performance of the (South Indian) Moplah Rifles on the North-West Frontier, he pitied their officers and supported the decision to disband them.

The selected martial classes

Service in the Indian Army in the late nineteenth and early twentieth centuries was, in the main restricted to the following groups:

Pathans from the North-West Frontier districts and the independent tribal territories;
Baluchis and Brahuis from Kalat and British Baluchistan;
Sikhs, Jats, Dogras, and Muslims from the Punjab;
Garhwalis, Kumaonis, and Gurkhas from the Himalayan region;
Rajputs, Brahmans, and Muslims from the Delhi and Hindustan regions;
Rajputs, Jats, Mers, and Muslims from Rajasthan and central India;
Marathas and Deccani Muslims from western India;
Christians, Untouchables, Tamils, and Muslims from southern India.

The Muslims (with the exception of converts in Bengal and parts of southern India) were generally descendants of military colonists, mercenaries, or settlers of Turkish, Afghan, or other foreign stock, and almost without exception all the major classes recruited, whether Muslim, Sikh, or Hindu, were men of yeoman or peasant status. These were men with the most inert political outlook, conservative, and willing to conform to the established order. They were also easier to command, for they knew their place, and their officers were not obliged to deal with the sort of judgement of their less reasonable orders that a quicker-witted or more intelligent and shrewder soldier would have made. Indeed when classes were cut from the recruiting list it was commonly those of the most independent turn of mind, such as the Pathans and Baluchis, or those best educated and most articulate, such as the Tamils, who were deemed to have 'difficult' or 'unreliable' ways.

Those selected by the British as 'martial' were in fact considered by the bulk of Indian opinion as rather bone-headed. One Punjabi proverb holds that three things are improved by beating; women, wheat, and a Jat (this very word means cultivator, and can be a pejorative term to the sophisticate, just as the words boor, pagan, churl, or peasant have become in English). A passage set for study in a well-known Urdu (or Hindustani) primer, although losing something in the translation, indicates the way in which the martial classes, who provide two out of the three characters in the story, appeared to ordinary townspeople:

> Once upon a time, a Pathan happened to go to Delhi. At the railway station he came across a Buniah and a Sikh with whom he made friends. All three started with a view to paying a visit to the place. Wandering about, they happened to reach the Jama Masjid. The Pathan on seeing it was very much astonished and said, 'How tall

the people must have been in former days, to make such high minarets!' The Sikh, on hearing this, said, 'No doubt the Pathans are fools, Listen, O foolish man, I will tell you how these were made! They first made them flat on the ground and then made them stand this way.' The Buniah, on hearing this, laughed, and said, 'Neither of you idiots knows how these were made. I am a Buniah. I am supposed to be the wisest and most cunning man in the world. I will tell you the right thing. They were wells first, but the people turned them upside down and now they are minarets.'

Tables 3–6 show the changes that occurred in the class composition of the Bengal Army between 1864 and 1884*. By the latter date nearly all the low-caste men had been mustered out, and recruiting from their communities had ceased. In 1865 five out of the nineteen cavalry regiments were mixed, eleven had single-class troops, and one was a single-class regiment. Of the infantry, eighteen battalions were mixed, sixteen had class companies and eleven were officially, or in practice, class battalions. By 1884 the cavalry was all organised into single-class troops, with three single-class regiments. The infantry, including the Punjab Frontier Force, had sixteen single-class battalions, and the others were all organised in single-class companies.

In 1892 further changes took place. The sixteen 'Hindustani' or non-Punjabi infantry regiments remaining in the Bengal Army were all reorganised as single-class regiments. Five groups were chosen, and the 1st and 3rd regiments became all Brahmans, the 2nd, 4th, 7th, 8th, 11th, 13th, and 16th all Rajputs, the 5th, 12th, 17th and 18th all Muslims (or Musalmans as they were called in India), the 6th and 10th all Jats, and the 9th all Gurkhas.

The class-company system, although it had the advantage of giving men of different religions or communities the opportunity of competing with each other in efficiency and thus, through their rivalry, keeping up the standards of the whole unit, had one major disadvantage. Men of one community could not be placed in command of those belonging to another. Thus whereas, in a single-class regiment, promotion could be according to a single seniority list, and soldiers or NCOs could be interposted from one company to another, in a class-company regiment this could not be done and separate rolls had to be kept for each of the communities concerned. If the Indian officers of one company became casualties, there was no way of replacing them by experienced officers from within the regiment.

An additional difficulty was caused by permitting regiments to recruit their own men. Although it formed a link between the British recruiting officers and the villages or clans from which they drew their men, it allowed the prejudices of individual regiments (generally shared by its officers and men alike) to restrict recruitment to an ever-narrower area. An incredible amount of time and effort was poured out to recruit men of a particular sub-class or district for a particular company. This meant that men even of the same community could not easily be transferred from one regiment to another because the second regiment entertained men of a different sub-caste or clan within that community.

The concentration of particular groups into special units or sub-units might also lead to complications arising from the local politics of the areas from which men were

* Adjutant-General's circular No 117 N. of 9 September 1864, and General Order of the Commander-in-Chief, 20 January 1883, cited in F. G. Cardew, *Services of the Bengal Army* (1903), pp. 329–31, 405–9.

Table 3

Class composition of Bengal Cavalry, 1864, showing proportion of class troops

Regiments	Multanis	Hindustani Muslims	Trans-Indus and Border tribes	Punjabi Muslims	Hindustani Hindus	Sikhs	Dogras and Hill-men	Bundelas	Jats	Total Troops	Remarks
1st Bengal Cavalry	The composition of the corps to remain as previously, viz entirely Hindustani Muslims.										
2nd " "	To be composed gradually, but eventually, of Muslims, Dogras, Sikhs, Jats, Rajputs, Brahmans, and Mahrattas, in equal numbers, and mixed together in the several troops.										
3rd " "	The same as 2nd Bengal Cavalry.										
4th " "	Ditto ditto										
5th " "	Ditto ditto										
6th " "	...	1	1	...	1	1	...	1	1	6	That is, class troops
7th " "	The same as 2nd Bengal Cavalry										
8th " "	...	1	1	1	1	...	...	1	1	6	
9th " "	...	...	1	2	...	2	1	...	...	6	
10th " "	...	...	1	1	...	2	1	...	1	6	
11th " "	...	...	1	1	...	3	1	...	...	6	
12th " "	...	...	1	1	...	2	½	½	1	6	
13th " "	...	...	1	2	...	2	1	...	...	6	
14th " "	To remain as formerly a class regiment of Jats only.										
15th " "	To remain as formerly a class regiment of Multanis, Duranis, Pathans, Baluchis etc.										
16th " "	...	1	1	...	2	...	1	...	1	6	
17th " "	...	1	1	1	1	...	1	...	1	6	
18th " "	...	1	...	1	1	1	...	...	2	6	
19th " "	1	...	1	1	...	2	½	...	½	6	

Table 4

Class composition of Bengal Infantry, 1864, showing proportion of class companies

Regiments	Brahmans and Rajputs	Hindustani Muslims	Jats	Gurkhas and Hillmen	Bundelas	Ahirs	Pasis	Lodhs	Dhanuks	Kurmis	Gujars	Chumars	Mehtars	Hindus of inferior castes	Punjabi Muslims	Trans-Indus and Border Tribes	Dogras and Hillmen	Trans-Sutlej Sikhs	Cis-Sutlej Sikhs	All races and castes
1st Native Infantry	To be recruited as formerly; a predominance of Rajputs with large numbers of Brahmans and Hindustani Muslims, some Sikhs and low caste men.																			
2nd ” ”	As formerly, chiefly of Hindustani Muslims, Brahmans and Rajputs, a few low caste men.																			
3rd ” ”	The same as the 2nd																			
4th ” ”	Do. do.																			
5th ” ”	2	1	1	1	1	...	...	...	...	...	...	...	...	1	...	...	...	1	...	...
6th ” ”	1	1	2	...	...	...	...	...	...	...	...	...	...	1	1		1	1	...	...
7th ” ”	As formerly, chiefly Hindustani Muslims, Brahmans and Rajputs, with an admixture of Cis-Sutlej Sikhs and low castes.																			
8th ” ”	2	1	...	...	...	...	...	...	...	...	...	...	...	1	...	1	1	1	...	1
9th ” ”	2	1	1	1	2	...	...	...	...	...	...	...	...	...	...	...	1	...	...	...
10th ” ”	2	1	1	...	1	...	...	...	...	...	...	...	...	1	1	...	1	...	...	...
11th ” ”	To be recruited as formerly, much the same as the 2nd.																			
12th ” ”	As the 2nd, with an admixture of Punjabi Muslims.																			
13th ” ”	1	1	2	1	2	...	...	...	...	...	...	...	...	1	...	...	...	...	...	...
14th ” ”	As formerly, principally Cis-Sutlej Sikhs, with a small admixture of Punjabi Muslims and Trans-Sutlej Sikhs.																			
15th ” ”	As the 14th.																			
16th ” ”	As formerly, of Hindustani Muslims, Brahmans and Rajputs, with a few Ahirs, Sikhs and low castes.																			
17th ” ”	As formerly; much the same as the 2nd.																			
18th ” ”	2	2	...	2	1	...	...	...	...	...	...	...	...	1	...	...	...	...	...	...
19th ” ”	To be composed of Punjabi Muslims and Trans-Sutlej Sikhs in nearly equal proportions, with a few Dogras.																			
20th ” ”																				
21st ” ”																				
22nd ” ”																				
23rd ” ”	A class regiment of Mazbi Sikhs.																			

Table 4 *cont*

Class composition of Bengal Infantry, 1864, showing proportion of class companies

	Rgeiments	Brahmans and Rajputs	Hindustani Muslims	Jats	Gurkhas and Hillmen	Bundelas	Ahirs	Pasis	Lodhs	Dhanuks	Kurmis	Gujars	Chumars	Mehtars	Hindus of inferior castes	Punjabi Muslims	Trans-Indus and Border Tribes	Dogras and Hillmen	Trans-Sutlej Sikhs	Cis-Sutlej Sikhs	All races and castes
	24th Native Infantry	The same as the 19th Native Infantry.																			
	25th " "																				
	26th " "																				
	27th " "																				
	28th " "																				
	29th " "																				
	30th " "																				
	31st " "	The same as the 19th Native Infantry but Cis-Sutlej Sikhs may be entertained.																			
	32nd " "	A class regiment of Mazbi Sikhs.																			
Low caste Levies	33rd " "	...	1	...	...	...	3	1	...	...	...	...	1	...	1	...	...	...	...	...	1
	34th " "	1	...	...	...	...	1	...	1	1	...	...	1	1	1	...	...	...	...	...	1
	35th " "	...	...	...	...	...	2	...	2	1	...	...	2	1	...	...	...	...	...	...	...
	36th " "	2	...	1	...	...	2	...	...	...	2	...	...	...	...	...	...	...	...	...	1
	37th " "	1	1	1	...	...	1	...	...	...	1	2	...	...	...	1				...	...
	38th " "	2	...	2	...	...	2	...	...	...	1	...	...	...	...	...	...	...	...	...	1
	39th " "	1	...	1	1	...	1	...	...	...	...	...	1		1	...	...	...	...	1	1
	40th " "	2	...	...	...	...	3	...	...	...	...	...	...	...	1	...	...	...	1	...	1
	41st " "	2	1	2	1	...	...	...	...	...	...	1	...	...	...	...	...	1	...	...	...
	42nd " "	Chiefly of Gurkhas and Hillmen (Assamese), with a proportion, not exceeding one-fourth of its strength, of Hindustanis.																			
	43rd " "	The same as the 42nd.																			
	44th " "																				
	45th " "	As formerly, to be composed entirely of Punjabis and Sikhs.																			
	1st Gurkha Regiment	Class regiments, entirely of Gurkhas and Hillmen.																			
	2nd " "																				
	3rd " "																				
	4th " "																				

Table 5

Class composition of Bengal Cavalry, 1884

Corps	Class regiment or class troop	Muslims			Hindus				Remarks
					Punjabi		Hindustani		
		Hindustani	Punjabi	Trans-Indus and Border	Sikhs	Dogras	Rajputs	Other Hindus	
1st Bengal Cavalry	C.R.	6	...	...	...	...	...	...	
2nd " "	C.T.	2	...	...	1	...	1	2(*a*)	(*a*) Jats.
3rd " "	C.T.	3	...	...	1	...	1	1(*b*)	(*b*) "
4th " "	C.T.	3	...	...	1	...	...	2(*c*)	(*c*) "
5th " "	C.T.	2	...	...	1	...	1	2(*d*)	(*d*) "
6th " "	C.T.	2	...	...	2	...	1	1(*e*)	(*e*) "
7th " "	C.T.	2	...	...	1	...	1	2(*f*)	(*f*) One troop of Jats and one of Brahmans
8th " "	C.T.	3	...	...	1	...	1	1	
9th " "	C.T.	...	2	1	2	1	...	...	
10th " Lancers	C.T.	...	1	1	2	2	...	...	
11th " "	C.T.	...	1	1(*g*)	3	1	...	...	(*g*) One-fourth of the troop might consist of Afridis
12th " Cavalry	C.T.	...	2	...	3	1	...	...	
13th " Lancers	C.T.	...	1	1	2	2	...	...	
14th " "	C.R.	...	...	...	...	...	...	6(*h*)	(*h*) Jats
15th " Cavalry	C.R.	...	2	4	...	...	...	...	
18th " "	C.T.	...	4	...	2	...	...	...	
19th " Lancers	C.T.	...	1	1	2	1	...	...	and one troop of independent trans-border tribes

Table 6

Class composition of Bengal Infantry, 1884

Corps	Class regiment or class company	Muslims				Hindus: Punjabi: Sikhs		Hindus: Punjabi	Hindus: Hindustani			North-East Frontier		Jarwahs of Assam	Remarks
		Hindustani	Punjabi	Independent Trans-border	Trans-Indus and Border	Jats, etc	Mazbis	Dogras	Brahmans	Rajputs	Other Hindus	Gurkhas	Hillmen		
1st Bengal Infantry	C.C.	2	...	...	...	...	...	...	2	3	1	...	...	...	
2nd ,, ,,	C.C.	2	...	...	...	...	...	...	2	3	1	...	...	...	
3rd ,, ,,	C.C.	2	...	...	...	...	...	...	3	2	1	...	...	...	
4th ,, ,,	C.C.	2	...	...	...	...	...	...	2	3	1	...	...	...	
5th ,, ,,	C.C.	2	...	...	...	...	...	...	1	2	3*(a)*	...	...	...	*(a)* Two of Jats
6th ,, ,,	C.C.	2	...	...	...	...	...	...	...	2	4*(b)*	...	...	...	*(b)* Ditto
7th ,, ,,	C.C.	2	...	...	...	...	...	...	1	3	2	...	...	...	
8th ,, ,,	C.C.	2	...	...	...	...	...	...	1	3	2	...	...	...	
9th ,, ,,	C.C.	1	...	...	...	...	...	...	1	2	2*(c)*	...	2	...	*(c)* Jats
10th ,, ,,	C.C.	2	...	...	...	...	...	...	1	2	3*(d)*	...	...	...	*(d)* Two of Jats
11th ,, ,,	C.C.	2	...	...	...	...	...	...	2	3	1	...	...	...	
12th ,, ,,	C.C.	3	...	...	...	...	...	...	1	2	2	...	...	...	
13th ,, ,,	C.C.	2	...	...	...	...	...	...	...	1	4*(e)*	...	1	...	*(e)* Two of Jats
14th ,, ,,	C.R.	...	1	...	...	7*(f)*	...	...	...	...	...	...	...	...	*(f)* Cis-Sutlej Sikhs chiefly
15th ,, ,,	C.R.	...	1	...	...	7*(g)*	...	...	...	...	...	...	...	...	*(g)* Ditto
16th ,, ,,	C.C.	2	...	...	...	...	...	...	1	3	2	...	...	...	

17th ” ”	C.C.	3	...	...	...	...	...	...	2	3	...	...	...	...	
18th ” ”	C.C.	2	...	...	...	...	...	...	1	2	1	...	2	...	
19th ” ”	C.C.	...	2	1	1	4(*h*)	...	...	...	...	...	...	...	...	(*h*) Cis-Sutlej Sikhs prohibited
20th ” ”	C.C.	...	1	2(*k*)	1	2(*l*)	...	2	...	...	...	...	...	...	(*k*) Afridis
															(*l*) Cis-Sutlej Sikhs prohibited
															(*m*) Afridis
21st ” ”	C.C.	...	2	1(*m*)	1	3(*n*)	...	1	...	...	...	...	...	...	(*n*) Cis-Sutlej Sikhs prohibited
22nd ” ”	C.C.	...	3	...	...	4(*o*)	...	1	...	...	...	...	...	...	(*o*) Ditto
23rd ” ”	C.R.	...	...	...	...	...	8	...	...	...	...	...	...	...	
24th ” ”	C.C.	...	2	...	1	3(*p*)	...	2	...	...	...	...	...	...	(*p*) Ditto
25th ” ”	C.C.	...	3	...	...	3(*q*)	...	2	...	...	...	...	...	...	(*q*) Ditto
26th ” ”	C.C.	...	...	1(*r*)	2	4(*s*)	...	1	...	...	...	...	...	...	(*s*) Ditto (*r*) Afridis
27th ” ”	C.C.	...	2	...	2	3(*t*)	...	1	...	...	...	...	...	...	(*t*) Ditto
28th ” ”	C.C.	...	2	...	2	3(*u*)	...	1	...	...	...	...	...	...	(*u*) Ditto
29th ” ”	C.C.	...	2	...	...	4(*w*)	...	2	...	...	...	...	...	...	(*w*) Ditto
30th ” ”	C.C.	...	2	...	...	4(*x*)	...	2	...	...	...	...	...	...	(*x*) Ditto
31st ” ”	C.C.	...	2	...	...	4(*y*)	...	2	...	...	...	...	...	...	(*y*) Cis and Trans-Sutlej Sikhs
32nd ” ”	C.R.	...	...	...	...	...	8	...	...	...	...	...	...	...	
33rd ” ”	C.C.	2	...	...	...	...	...	...	2	2	2(*z*)	...	...	...	(*z*) Jats etc
38th ” ”	C.C.	...	...	...	...	...	...	...	...	2	4(*aa*)	...	2	...	(*aa*) 2 of Jats
39th ” ”	C.C.	...	...	...	...	...	...	...	1	1	4(*aa*)	...	2		
40th ” ”	C.C.	...	...	...	...	...	...	...	1	3	4	...	...	...	
42nd ” ”	C.R.	...	...	...	...	...	...	...	...	...	...	7	...	1	
43rd ” ”	C.R.	...	...	...	...	...	...	...	...	...	...	7	...	1	
44th ” ”	C.R.	...	...	...	...	...	...	...	...	...	...	7	...	1	
45th ” ”	C.R.	1	...	...	...	6	...	1	...	...	...	...	...	...	
1st Gurkha Regiment 2nd ” ” 3rd ” ” 4th ” ”	C.R.	...	...	...	...	...	...	...	...	...	...	8	...	...	

drawn. The 40th (Punjab) Regiment of Bengal Infantry was recruited mainly from Pathans of the independent trans-frontier tribal territories. Indeed in the Kitchener reorganisation its title was changed to that of the 40th Pathans. The unfortunate coincidence of that number with the allegedly dishonest propensities of its men led to this regiment being known unofficially as the Forty Thieves, its colonel (or any of its officers attached to another regiment) receiving the sobriquet of Ali Baba, and its adjutant, that of Abu Hasan, who was, in the story, the 'Captain of the Robber Band'. In the lawless state of much tribal territory, clan feuds and private vendettas flourished. If several men of one faction were killed in the British service, they left their own people at a disadvantage in local fights, and therefore the prudent Pathan would not expose himself to danger too rashly in his employers' quarrel on that account.

Some British officers were so obsessed by the virtue of the particular class from which they chose to recruit that they would hear nothing to the advantage of others. They became partisans in the rivalry of their men, and even, like the officers of the old Bengal Army, entered into their caste feelings. Some British officers would not countenance organised games between their high-caste troops and men of lower caste. Such officers, by over-identifying with one particular class of troops, virtually unfitted themselves for higher command, since the authorities, although generally in favour of the concept of martial classes as such, could not accept that any one of them was better than another.

Indeed, it was the overall policy to emphasise the differences between the selected martial classes, and so to create a rivalry in performance, together with a parity of esteem. The different regions and religions of India produce different types of men, as do those of Europe or North America. The British Indian Army systematised them in its soldiers. The special virtues of each class, some real, others illusory, were developed and exploited to improve regimental *esprit de corps*. Clan or class loyalty was intensified by identification with the regiment or company. Sometimes this was a self-reinforcing process, as men returned to their villages with their turbans tied or beards worn in the regulation manner, adopted by their regiments for the sake of uniformity, and retained these fashions in their civil dress. They and their families kept up these distinctions, to emphasise their military connections and out of pride that they were of a community that had the privilege of performing military service. Later generations, Indian as well as British, would come to consider these fashions as caste distinctions, sanctified by tradition, and so insist that the regiment was bound to preserve in its uniform the distinctive dress of the caste from which it was recruited.

The Sikhs

Regiments which consisted wholly or partly of Sikhs were the most zealous in preserving distinctions of outward appearance, which indeed really were prescribed by religious authorities. Under persecution from the Mughal Emperor Aurangzeb, a Muslim zealous in the faith, the Sikhs had been driven to adopt warlike practices. They formed a military fraternity, the *Khālsa* or Elect. Its members took the name *Singh* or Lion. They abjured wine and tobacco, ceased to shave their beards and hair, and wore five objects beginning with the letter *K*, long hair, drawers, a comb, a dagger, and an iron discus (*kes*, *kaccha*, *kankan*, *kirpan*, *kangha*).

As the Mughal power declined the Sikhs returned from the hills, deserts, and jungles where they had been driven, and re-established themselves in their Punjab homelands. Under Ranjit Singh a powerful state was built up in the early years of the nineteenth century, with its own regular army. The Sikhs were by now accustomed to consider the military life a normal, even a desirable existence and they were mostly from the sturdy agricultural classes of the Punjab, physically tough and able to acquit themselves well in battle.

During the Sikh Wars of the 1840s these men impressed their British opponents with their toughness and martial spirit. Many were recruited into the Indian Army after the annexation of the Punjab. During the Indian Mutiny most Sikhs supported the British cause, and in flocking to aid the forces before Delhi were able to combine the chance to loot a great city with the opportunity of repaying old scores against the last remnants of their Mughal persecutors.

In the service of the British the Sikhs continued to wear the five Ks in their military uniform. The *Granth Sahib*, the Sikh bible, was accorded official respect and carried at the head of all Sikh regiments on the line of march or placed in a prominent position in camp, rather like the Ark of the Covenant in the host of the Israelites. Men who lapsed from their religious observances could expect to incur the disapproval of their military as much as their spiritual leaders. Indeed, such was the measure of British military support for the Sikh religion, that it has been suggested that this foreign influence was a major faction in preserving its separateness and preventing it from being absorbed into the main body of Hinduism as just another sect.

This encouragement of Sikhism was not without its inconveniences, or even its dangers. When steel helmets were issued during World War I many Sikhs refused to wear them in place of the large turbans which held their uncut hair. In vain did their officers point out to them that the *pagri*, as such, was not prescribed by religion, and that the soldiers of Ranjit Singh had not scrupled to wear iron caps. The Sikhs had come to associate their uniform *pagri* with their religion and would not be persuaded otherwise. The Akalis, or zealots, emerged as a political reformist movement in the early years of the twentieth century, and attracted many Sikh soldiers after World War I. Very great alarm was felt in official circles when whole villages sympathised with the Akalis, and recruiting from affected areas ceased for several years. Restrictions on Indian immigration to Canada in 1914 led to much Sikh unrest, several retired sepoys destroying their discharge papers in sympathy with their excluded co-religionists. In the 1930s the emergence of shaven Sikhs, men who no longer wished to observe the letter of the religious law, puzzled the Army. Eventually they were accepted, in the light of contemporary military changes, as 'mechanised Sikhs' and all was well.

Punjabis

Other Punjabis, not Sikhs by religion but of the same ethnic stock, were held by the British in equally high esteem as good military material. Punjabi Musulmans (Muslims) and Jats and Dogras (Hindus) were equally sought after and given a high place in the list of martial classes. When most of the Madras and Bombay regiments adopted the class-company system in the early 1890s the opportunity was used to form companies from Punjabis instead of local men.

In 1862, after the post-Mutiny reorganisation, the Indian infantry consisted of 28 battalions recruited predominantly from the Punjab, 5 from the Gurkhas, 28 from Hindustan, 30 from Bombay and 40 from Madras. Thirty years later there were 34 from the Punjab, 15 from the Gurkhas, 15 from Hindustan, 26 from Bombay and 25 from Madras. Just prior to Kitchener's reforms in 1902, out of 30 battalions in the Madras line 14 were entirely recruited from Punjabis. These were the 2nd, 6th, 7th, 9th, 12th, 14th, 22nd, 24th and 27th Madras Infantry, and the 29th, 30th, 31st, 32nd and 33rd Burma Infantry. In the Bombay infantry 16 out of 26 battalions had two or more companies (out of eight) formed from Punjabis. After the Kitchener reforms the Punjabisation of the Indian Army went even further. On the outbreak of war in 1914 there were 57 battalions of Punjabis, 15 from Hindustan, 18 from Bombay, and 20 from Madras, most of whom were recruited from Tamils or down-country Muslims. Telugus, who had made up 25 per cent of the Madras Army at its peak, were no longer recruited at all.

Henry Lawrence, the distinguished Indian statesman and soldier, one of the three original members of the Commission that governed the newly conquered Punjab in the late 1840s, wrote that 'Courage goes much by opinion; and many a man behaves as a hero or a coward, according as he considers he is expected to behave. Once two Roman Legions held Britain; now as many Britons might hold Italy.' It was a hundred years before this view was accepted in senior military circles, and a memorandum on recruiting written during World War II officially condemned 'parochial minded' British and Indian officers, 'vested class interest' and 'bogus caste prejudice' for maintaining the Army on a narrow and inefficient system of recruitment.

The martial-caste system was flattering to the British, who were able to see themselves as the supreme caste, able to command all races. It even went far to reinforce, in military men at least, the existing social and class prejudices of their own, British, society. One official Government of India publication, *Our Indian Empire*, containing hints for the use of British soldiers proceeding to India, when discussing the martial races of India says: 'The Gurkha has no more in common with a bazaar shopkeeper than an Englishman with a Jew pedlar.' This, which seems to deny that an Englishman can be either a Jew or a pedlar, was published in 1935, and could easily have come from a contemporary official German publication.

The motivation of the soldiers

There was no conscription or national service system in force under the British, and every man who joined was, as in the British Regular Army, a free volunteer. Many of the reasons why an Indian soldier enlisted were those for which young men have always enlisted, at other times, and in other countries. These included the glamour of military clothing, shining and glittering accoutrements which set him apart from and above the mass of toiling humanity, and which are thought to find him favour in the eyes of young women; the chance of adventure in strange places; the chance to win glory and fame; and the pride of bearing arms, a privilege which is not granted to all men (nor, for that matter, desired by all men, although the soldier might not see this). There might also be a family tradition of military service, and a desire in the young man to follow in the steps of his ancestors and emulate their deeds, or pressure on him to continue the line of service that had brought esteem to his family. These factors are not found among Indian soldiers alone, but were especially strong in a society where family ties were so close-knit, and where the caste system emphasised both the right and the duty of a man to take up the employment followed by his forefathers.

Indian officers were keen to recruit men from their own clans to build up their position in the regiment, and were themselves under pressure from their families to find jobs for the young men reaching military age. Leaders of communities, too, put pressure on young men to join, to ensure that they kept their social status as a military caste, and the local magnate who produced the required quota of recruits could count on receiving awards and decorations from a grateful government. One who failed would receive lectures and veiled threats from touring officers, pointing to the favourable response of other communities, and to the eagerness of other classes to be allowed to fill any vacancies.

In time of scarcity, too, the Army offered a secure career, and a chance for the younger son to escape from charges of loitering about the farm in idleness. So the warm coat and the kind word from the squire were incentives to the Indian as much as to the British farm lad to join the Army and seek his fortune. But in India, as in Britain, times of prosperity in the countryside were thin times for the recruiting officers. The Bombay Army especially lost many potential warriors to well-paid employment in the growing industrial areas of western India. The prosperous farmer, when food prices rose (as they tended to when governments bought grain in wartime), preferred to stay at home and keep his sons with him, no matter how anxious the regiments might be to obtain their services.

There was no patriotic motive to join. India was not a unified country prior to the British conquest, and many served, in the early days of British rule in India, who were not, strictly speaking, British subjects. Even after British control was complete, and the British monarchy assumed the title *Kaisar-i-Hind*, the ordinary recruit was not impelled by the idea of service to his country. He may have felt loyalty to a distant white ruler, but it was generally as a client or an associate of an alien ruler that he saw his place in society.

It was considered quite respectable to be the agent of an alien paymaster. Indian soldiers had for generations served whoever held power in the State, and indeed whoever was sufficiently well organised and endowed to try and seize it. The Indian soldier was, then, a mercenary, in that he served in return for pay and other material or moral rewards. He was not, however, a hireling who sold his services to the highest bidder; on the contrary, as long as his employers kept faith with him, he kept faith with them.

An Indian soldier's story

Only one work has so far been published in English in which an Indian soldier tells his own story of life in the ranks in the first half of the nineteenth century. This is *From Sepoy to Subedar* by Sita Ram Pande, originally translated and published by Lieutenant-Colonel J. T. Norgate in Lahore in 1863, and republished at intervals thereafter, the Urdu version being used as an official textbook for British officers studying for the Higher Standard Urdu examination in both world wars. The most recent edition, by General James Lunt (Routledge & Kegan Paul, 1970), discusses in some detail the authenticity of the memoirs, and comes to the conclusion that although doubtless embroidered or mis-remembered by Sita Ram himself (who would have been about seventy years old at the time of first publication) and amended by subsequent editors, the work is, in essence, a reliable source of information.

Sita Ram's career, which lasted from 1812 to 1860, was one in which many features common to other Indian soldiers are to be found.

He came from an agricultural village, in Awadh or Oudh. His family held about 150 acres of land, and were of the Brahman class. He was thus a high-class Hindustani or 'Purbia', the group that was the most favoured 'martial caste' before the Indian Mutiny, just as the Punjabis were to become afterwards, and he was, too, a scion of the 'yeoman stock' which the British preferred to recruit. He was attracted to the Army by the tales told by a serving 'old soldier', by the fine appearance of his uniform (a glittering scarlet coat with brass buttons) and by the ready money which he freely spent on furlough. These factors have much the same influence in persuading young men the world over to join up, but in Sita Ram's case there was an added influence in that the 'old soldier' was his own uncle, a jemadar or Indian lieutenant of infantry, and we have already noted that the Indian officers of a regiment recruited men from their own families as much as possible to build up their own status and following in the regiment. Later, after the death in action of his uncle and many of his friends, Sita Ram was to leave his old corps, judging rightly that his prospects would then be better in another newly raised regiment.

Sita Ram's mother, and his tutor, the family priest, were both horrified to hear of his decision to become a soldier, declaring that the stories of army life were untrue, that the unpleasant aspects of it had been glossed over, that he would be degraded by military life, that he would be killed, or maimed, that his education would all be wasted, and so on. This, too, is typical of the mothers and teachers of would-be soldiers in many other countries. It is, however, interesting to see that military service under the Government was by no means held in such universal esteem as many military writers would suggest. Sita Ram's father agreed, however, as he had a lawsuit outstanding, and his son's appointment as a soldier of the East India Company would ensure that the claim would be given prompt attention by the King of Oudh, since it would be forwarded as a petition through his commanding officer to the British Resident at Lucknow. This is one instance of the way in which there were incentives for men who wished to prosper under British rule to supply soldiers to the Indian Army.

Sita Ram was actually under age on enlistment, as the regulations had since 1796 stated that recruits were to be accepted only between the ages of sixteen and thirty. Sita Ram was in his fifteenth year, but this would be near enough for any recruiting officer, and he presumably met the height requirement of five feet six inches, as he was accepted without difficulty. His term of enlistment would have been for three years initially, after which soldiers could normally claim to be discharged on two months' notice. There was no discharge in time of war.

On joining the Army, he was at first the new raw recruit up from the country, astonished at the new sights and sounds of the barracks, awed by his colonel and adjutant, baffled by the medical inspection, and bullied by the drill sergeants, so that, like many another recruit, he began to think better of his decision to be a soldier. However, after eight months' training he passed off the square, ahead of the rest of his squad, and was judged fit to take his place in the line of the 2nd Battalion of the 26th Bengal Native Infantry.

As an infantryman he found himself carrying a load, made up of his musket, pack, cross-belts, bayonet and cartridge-box that no respectable person would dream of carrying

The last stand of the 44th Foot on the retreat from Kabul, January 1842. Note the body of the Bengal Horse artilleryman in the right foreground. Oil painting by W. B. Wollen

The 2nd Battalion, Prince of Wales's North Staffordshire Regiment, Dagshai, Punjab. Photograph 1904

in the Indian heat if he could afford a porter. He also found the scarlet coatee too tight under the arms and the head-dress too hard and too heavy, although in this case he seems to have resigned himself to the fact that young soldiers, like young women, must suffer to be beautiful. Western-style uniforms, even when considerably modified after the Indian Mutiny, were always a trial to the soldier, and many regiments invented uniforms of Indian civilian pattern, called regimental *mufti*, for wear on occasions when military duties were not being performed. Sita Ram remarks how he always found it a relief to wear the loose-fitting cotton garments normal in his own country, and while it is usually more comfortable to be in loose civilian clothes rather than tight or heavy military ones, the temperature of the Indian plains must have made constrictions imposed by military fashion so much the more unpleasant. He does not complain about his footwear, but this is because at this time, and indeed until almost the end of the century, Indian infantrymen wore their own shoes or sandals of the local style. At least two major reviews in the 1870s were ruined by large numbers of soldiers, marching past on muddy ground after cavalry, stopping or breaking ranks to pick up their shoes which had come off in the mud, and which, being their own property, could not be left behind. In later years, the issue of leather ammunition-boots, heavy, laced up to the ankles, and with steel tips, gave a securer grip, but were a sore trial to boys from districts where such restricting footwear was unknown.

Sita Ram tells how he served in the Pindari campaign, in the First Afghan War – where he was taken prisoner during the retreat from Kabul – during the Sikh Wars, and finally during the Indian Mutiny. He was wounded seven times, but the medical arrangements seem to have been satisfactory, as in most cases he was carried away in a doolie, or litter, by bearer parties, to the hospital tents, and there treated until sufficiently recovered to go on sick leave.

The Indian soldier, if he was sick or wounded, was (until after World War I), treated in his own regimental hospital. This consisted of huts or tents, or large bungalows, presided over by a European medical officer and an Indian assistant, with orderlies from the regimental soldiers, and 'followers' who attended to domestic or menial tasks. There were a few supplies, some drugs and surgical instruments, beds, mattresses, pillows and blankets, but very little else. Patients brought their own bedding and clothing wherever possible, and supplemented official rations with their own food. The different religions and classes of the patients made for difficulties in the case of mixed regiments, and in such cases it was easier to let men be looked after by their own colleagues than attempt to centralise diets or feeding. As soon as the physician or surgeon had done his work, men were discharged on leave.

Sita Ram, as a high-class Hindu, found his religious observances especially onerous when wounded or a prisoner. On two occasions he had to undergo expensive purification ceremonies to regain his place in his class after being defiled by contact with lower-class Hindus or Muslims.

The levitical scrupulousness of Brahmans in this respect tended to be a cause of annoyance to the British officers. There were numerous cases where men returning from campaigns or captivity were ostracised by their fellows, who would indeed be polluted themselves by contact with one who had become ritually unclean. Soldiers complained that the cost of the purification ceremonies coincided with the amount of back-pay owing to them, and this led their officers to hold the priests guilty of mercenary behaviour, tailoring their charges to take all the soldier had. Priests, as a group, were held in suspicion or dislike by many British officers, Christian ministers almost as much as Hindu priests or Muslim religious leaders, though the divergence of views between the spiritual and secular leaders of men was probably no more marked in India than in other countries. Nevertheless it is a paradox that at the same time as class and caste prejudices were most respected, the religions that gave rise to them were most despised.

Sita Ram in due course rose to the rank of havildar and became pay-havildar. The sepoys left their money with him until it was required, usually when going on furlough, and Sita Ram, following the custom of the service, lent it out at interest, usually to the British officers of his regiment who were perennially in debt. This was contrary to the regulations, and Sita Ram, having over-extended credit to one officer, was discovered and deprived of a lucrative source of income. He complains that although the Indian officers who tried him did not consider he had done anything wrong, they found him guilty since they believed that this was what their commanding officer wished to happen. He was not deprived of his rank, in view of his previous good record.

Sita Ram was appointed jemadar after the First Sikh War, having then served thirty-five years in the ranks and reached the age of fifty. He was by now an old man by Indian standards, and physically unsuited by any standards to perform the duties of an infantry subaltern. However, surviving the Mutiny, he was promoted to subedar, or Indian captain, at the age of sixty-five. He had thus gained the summit of his ambition, but was quite unable to do the duties which as a younger man he would have been well fitted to perform. He was therefore invalided out, but was granted the pension for which he had served so long, and ended his days in his own village, telling stories of his career to incredulous youths, as is the way of old soldiers.

Rates of pay

The financial rewards for which Sita Ram and his comrades served varied over the years. Pay was normally monthly, in arrears. Silladar cavalry were paid the highest rates, as they had to feed and equip themselves and their horses. The rates per month in 1861 for an irregular cavalry regiment were:

3 ressaldars, at Rs 300, 250, and 200
3 resseidars, at Rs 150, 135, and 120
(These were the six troop commanders, who were paid according to their relative seniority.)
1 wurdi-major (Indian adjutant) at Rs 130
6 jemadars, 2 each at Rs 80, 70 and 60
6 kot-dafadars or troop sergeant-majors at Rs 47
48 dafadars at Rs 38
6 nishanbardars (NCOs carrying the guidons) at Rs 38
6 trumpeters at Rs 34
420 sowars or troopers at Rs 27.

In 1891 pay was increased to Rs 31 per month for troopers and proportionately for the higher ranks. However, the cost to each man of maintaining himself, horse, and transport, was such that after deductions a soldier's 'take-home pay' was between Rs 8 and Rs 7 per month.

Regular infantry were paid at about one-third of these rates. Private soldiers earned Rs 7 per month until 1895 when the rate was increased to Rs 9. Non-silladar cavalrymen, and artillerymen, were paid Rs 2 and Rs 1, respectively, more than the infantry. An increase in 1877 improved the position of the Indian officers of infantry a little, the pay of the four senior subedars in each unit being raised to Rs 100 per month and the four juniors to Rs 80, while the senior and junior jemadars were paid Rs 50 and 40 respectively.

These rates of pay must be compared with those in employments outside the army. Rates varies a great deal throughout the period under review, and according to the part of India studied. Some idea of the average may be gained from the following figures, for 1831. In Calcutta, the capital of British India, skilled artificers, such as metalworkers or woodworkers, commanded a wage of between Rs 10 (£1) and Rs 12 per month. Unskilled artisans received about Rs 5 or 6, ordinary labourers about Rs 3·8, and agricultural workers anything between Rs 2 and Rs 6. In the smaller towns these rates were lower, and in country areas they were lower still. There a ploughman earned about 10 annas per month, and a labourer about 14. The price of rice, the staple grain, was at that time about Rs 1·8 per *mānd*, then equivalent to about a halfpenny per lb.

The sepoy, like the worker in civilian life, had to pay out of his wages the cost of his food, clothing, and quarters. Sepoys usually lived in huts, containing two or three men, built by them at their own expense. Special areas were allotted for the sepoys to build their huts, but these were without services such as water supply, drainage, or sanitation. Such areas were often pestilential and foul, and harboured diseases of many kinds. They were also fire traps, being crowded together and made with timber frames, thatched with straw or similar combustible materials. Eventually, barracks were built for Indian troops, from the end of the nineteenth century onwards. These were subdivided into cubicles each housing two or three men, rather than the communal dormitory system of the European soldiers. In peacetime most regiments, British or Indian, were based in 'cantonments', self-contained military townships, usually on the outskirts of the larger towns or cities.

Pensions were payable either in cash or in land grants. In Indian law it was not the land itself that was owned or disposed of, but rather the land revenue, the right to collect the government's share of the crop. Thus land was *held* rather than *owned*. The extent of the grant, or *jagīr*, varied according to rank. In the first part of the nineteenth century a subedar was granted 400 *Bīghās* (a *Bīghā* is a measure of land approximating to five-eighths of an acre) and a sepoy eighty *Bīghās*. The land was found from uncultivated or abandoned areas, of which there was a great deal at first in consequence of the disorders following the collapse of the Mughal Empire, but which eventually became less easy to find. If a *jagīr* were granted, the soldier's pension was reduced by one-third for the first three years of retirement, and after the next two years by another third. The remaining one-third of full pension was paid to him for life, and thereafter his heirs or assignees held the *jagīr* in perpetuity in return for a fixed annual fee, a quasi-rent, to the Government.

Pensions were payable after forty years' service, until 1886 when the qualifying period was reduced to twenty-one years. At the same time the invalid pension, paid to men discharged as unfit for service after fifteen years' service, was abolished and replaced by a bounty of one year's full pay to any soldier invalided out with between fifteen and twenty-one years' service. Soldiers discharged as unfit through wounds or injuries sustained in their duties also received pensions. A special feature of the pension system was that of a dependant's pension, payable to the next of kin, either widow, child, or parent of a soldier killed on active service. Another variant was that pensions could be awarded for two 'lives', ie, that of the soldier and then that of his heir. This was especially valuable in a country where the mortality rate from tropical diseases was so high that a man might not long survive to draw the pension he had earned.

In addition to his pay the soldier had extra emoluments. In 1876 every recruit for the artillery, infantry or sappers and miners was given a grant of Rs 30 for the purchase of personal items of clothing, and a further Rs 4 annually for its upkeep. In 1880 a bounty of Rs 50 was paid to all recruits, Rs 25 on enlistment and the rest on completion of three years' service. In 1886 it was decided that recruits should be paid from the date of enlistment rather than the date of actually joining their regiments, and would be granted free carriage of their goods from their homes to their station, or the field service allowance, *batta*, for the period of their journey. Good-conduct pay was instituted in 1837. Any sepoy with a clean crime sheet during the previous two years was paid an extra R1 per month after sixteen years' service, or Rs 2 after twenty. This was subsequently altered at different times, and by 1886 the soldiers could earn Rs 1, 2, or 3 per month after three, six, and ten years' service respectively. Prior to the General Service Enlistment Order of 1856, sepoys could also expect bounties for volunteering for overseas duty, which was one reason for the unpopularity of that order, which required every sepoy to accept, on enlistment, liability for foreign service.

Honours, awards and privileges

In 1837 two orders were instituted to reward, with both honour and ready money, the Indian officers and soldiers for their services. The Order of British India consisted of two classes. The first class was made up of subedars and above. They received an increment of Rs 2 per month to their pay, and were accorded the title of *Sardār Bahādur* (literally, the chief, the brave one). The second class was made up of Indian officers receiving a monthly increment of R1 and holding the title *Bahādur*. The Indian Order of Merit was awarded to any Indian officer or soldier, regardless of rank, for any act of conspicuous gallantry in the face of the enemy. One act of bravery gained admission to the third class, and an increase of one-third in pay and pension. The second act brought admission to the second class, with an increase equal to two-thirds of basic pay, and a third act was rewarded with the first class, and double pay or pension.

Until 1912, the Indian Order of Merit was the Indian soldier's VC. The Victoria Cross, instituted in 1856 as Britain's premier decoration for valour, could be, and was, awarded to European officers and men, and even civilians, in the service of the Government of India, but Indian personnel were excluded until 1912. The first Indian soldier to win this coveted award was Sepoy Khudadad Khan, a machine-gunner of the 129th Baluchis, for valour against the Germans on the Western Front, 30 October 1914. He survived the war and retired as subedar-major, the highest rank attainable by an ordinary soldier. Indian officers on retirement were occasionally awarded honorary commissions as captains or lieutenants. This was not only a signal honour, and appreciated as such, but also doubled the ordinary pension of an Indian officer, which was, in 1900, about Rs 10 per month.

The position of married soldiers actually serving with the colours varied between armies. In the Bombay Army, the arrangements were much the same as in the British. A certain proportion of families were accepted 'on the strength' of the regiment, and space for married quarters was provided for them within the regiment's lines. In Madras, all heads of families were allowed to bring their wives and children with them, but at their own expense. Roberts, when Commander-in-Chief, Madras, tried to cut down on the numbers of families actually with troops, as he considered them to be a drag on the military efficiency and martial spirit of the soldiers. The Bengal Army had no provision for married quarters. The soldiers were expected to leave their families in their villages, but frequent leave was granted to enable them to go home. The large 'joint family' system of Indian society enabled the wives and children of soldiers serving away from home to be protected and their property to be cared for by relatives.

Leave was usually granted on a generous scale. On average a soldier could expect two months' leave every year, plus a further six months or so every third year. Leave in such amounts was in fact quite necessary, at least before the coming of the railways, since men might have to walk up to 1,000 miles from a remote frontier station to reach their home village, and then make the return journey at the end of their furlough. For many men, it was necessary to be able to return to their villages and families, in order to take part in various festivals or family ceremonies, weddings, funerals, and the like, as dictated by their religion.

For those with the regiments who were not on leave, religious and regimental festivals of various sorts, designed to build up *esprit de corps*, could also be counted as holidays. On an ordinary working day, training started early, at first light while it was still cool, and finished before the noon-day heat. During the hot weather and the monsoons little training could be done, and, in the cavalry, the time of the troops was almost entirely taken up in exercising the horses.

The followers

No account of the Indian Army would be complete without mention of its followers, that vast separate army of non-combatants who lived with it in barracks and accompanied it on campaigns. Some were completely unofficial, the traders and stall-keepers who, attracted by the prospect of profit from business with men regularly paid in ready cash, set themselves up near to the garrison as soon as a military base was established. Sometimes such traders formed a regular regimental bazaar, unofficially attached to the regiment it served, and moving about the country with it as it was posted from place to place, even following it on active service, risking unpleasant consequences in the event of a sudden reverse or disaster. Other followers were menials of different sorts either employed as personal servants or in similar domestic capacities, such as mess cooks and waiters, bearers, native butlers and so on.

Official followers, authorised by headquarters, carried out a variety of tasks. Lascars were allotted for the care of tents, and to carry out tasks requiring organised and disciplined labour. Sweepers dealt with the disposal of waste, latrines, and refuse. The vital importance of water in Indian conditions was recognised by the appointment of *pakhālis* (water carriers with pack-bullocks), and *bihiṣtīs*, or bhistis (water-carriers with goat-skin bags) to distribute water to the troops. Many instances were recorded of the bravery of water-carriers bringing water to troops in the firing line. They were, strictly speaking, non-combatants, and had no means of defending themselves when attacked, yet carried out their duties even when exposed to enemy fire. The story of Kipling's famous poem 'Gunga Din' resembles many similar incidents in real life. The *Sāīs* (syce) or groom was an essential part of the support to mounted units, as were the grass-cutters. These men were not mere reapers, but had special skills and experience gained over many years. What to the untrained eye looked like a barren desert often yielded ample quantities of green-stuffs after the grass-cutters, knowing where and what to look for, had done their work. Cavalry were allowed one cutler per troop, to sharpen and maintain swords, lances, and other steel implements. In the Central India Horse a Sikh ressaldar fined his cutler, or sigligar, a rupee, because, in a fight near Lucknow during the Mutiny, his sword had failed to cut an adversary completely in half. The ressaldar declined to accept the argument of his British squadron officer that as the adversary had at least been completely killed, the sigligar should not be considered at fault.

Although not soldiers either by training or tradition, such followers were often as closely associated with their regiments as the fighting men, and son followed father, or nephew uncle, in much the same way. The martial-class theory and the custom of the country forbade them to bear arms, yet in an emergency brave men did so, and were all the braver for want of training as soldiers. Sweeper Itarsi, of the 125th Napier's Rifles, when the regiment suffered severe casualties in the battle of Sannaiyat in 1916, snatched up a rifle and bandolier and fought in the front line until the battle was over. He was afterwards appointed by the other sweepers to be their officer, and treated by them as such.

The soldier and the public

Indian public opinion varied in its estimation of soldiers much as did public opinion in other countries. Unsophisticated folk were impressed by tales of adventure and strange experiences, and so held soldiers in esteem on that account. Soldiers individually were on the whole well treated. Soldiers collectively were objects of suspicion, on account of the traditional military practices of making too free with other people's property, including their girls. The Indian aristocracy were generous in their attitude to the *jawāns*. (*Jawān* is the usual Urdu term for an Indian soldier. Strictly translated it means a youth, or lad, but like the Latin *juvens* to which it is cognate can mean, by extension, any man of military age and actually serving.) The soldier was an upholder of the established order, and the exponent of the military virtues of physical courage and skill-at-arms, virtues to which the aristocrat himself aspired. The Western-educated classes, while tending to regard the soldier as a hireling of the imperialists paid to conquer and oppress their fellow Indians, nevertheless saw in their martial exploits just cause for national pride. In both world wars nationalist newspapers carried stories of the valour of the *jawāns* against foreign enemies, and praised their achievements in battle.

Courage in battle

Of the courage shown by Indian soldiers only a few examples can be given. One hero was Jemadar Durga Singh of the Scinde Irregular Horse, on the border with Kalat, in 1850. This officer, on hearing that some camels had been stolen near his post, set out with twenty men and found their tracks. Thirty miles later, having lost seven horses through exhaustion, he came up with the robbers, who abandoned the camels and fled. The jemadar followed, but lost more and more horses until he had only a Baluch guide or tracker and two of his own sowars with him. The guide, observing that the fugitives had been joined by a band of about fifty of their fellow tribesmen, advised breaking off the pursuit. Jemadar Durga Singh replied that he would not return and report to his commander, Colonel John Jacob, that he had fled from an armed enemy. He and the two sowars then charged the enemy and died fighting.

At Maizar, on the Punjab frontier, in June 1897, a division of the 6th Battery of Indian Mountain Artillery was part of the escort to a Political Officer visiting a Pathan village. The tribesmen invited the British officers to a feast, where they were treacherously attacked and murdered. The infantry of the escort, a detachment of the 1st Sikhs, and the mountain gunners, were surrounded by swarms of fanatical tribesmen. The guns were brought into action, and, under command of the havildar-major, did considerable execution. Eventually, their ammunition expended, the gunners were obliged to withdraw. Despite losing many gun mules, they, and the infantry, returned safely to the main body of the Army, bringing with them their wounded, their guns, and their honour. Nine men won the Indian Order of Merit that day.

The heavy fighting in France and Flanders at the beginning of World War I gave the men from India many opportunities to display their courage. Fighting in a completely strange environment, far from home, cold, wet, and subjected to a far more heavily armed enemy than they could have ever imagined, the men of the Indian Corps in the British Expeditionary Force displayed courage of the highest order. Their courage was all the more remarkable in view of the appalling casualties among their British officers with whom the *jawāns* had established bonds of loyalty of the most personal kind. One incident alone must suffice as an example. At the battle of Givenchy, December 1914, a platoon of 59th Scinde Rifles (a regiment which despite its name was actually one of those in the former Punjab Frontier Force) got into a German trench and drove out its occupants. During this action their British officer, Lieutenant W. Bruce, was killed. They held on to the position all day until their rifles began to jam and the enemy brought up a trench mortar. Havildar Dost Muhammad, who had taken command on Bruce's death, ordered the survivors to retire. They refused, saying that their dead Sahib, Lieutenant Bruce, had ordered them to hold out to the end. They fought until all were killed except Havildar Dost Muhammad and one wounded man. These two then crawled back to the British lines, the havildar later being awarded the Indian Order of Merit.

The most famous of all such acts of loyalty was the defence of the British Residency at Kabul, 3 September 1879. In the Residency were the British Envoy, his secretary, one medical officer and one army officer, commanding an escort of twenty-five cavalry and fifty infantry of the Guides, part of the Punjab Frontier Force. They were attacked by overwhelming numbers of mutinous Afghan troops. Three times the little garrison sallied out, led by the Europeans. The fourth sally was led by a Sikh jemadar. The Afghans, realising that all the British were dead, offered the survivors the chance to come out and join them. The offer was refused with scorn, and the men of the Guides fought on until all were killed. The whole escort were posthumously admitted to the Indian Order of Merit.

These are only a few of the many cases where Indian soldiers died bravely and honourably, serving the rulers of their country, and keeping faith with their oath of loyalty and obedience.

7

The Officers

There were two groups of officers serving in India. One group belonged to the British Army, serving for a limited time there, either with their regiments or as individuals in particular appointments. The other group belonged to the Indian Service, employed by the Government of India, and permanently stationed there.

Officers of the British Army did not expect to be able to live on their pay. The expenses which they were normally required to meet in their social life were generally considerably greater than their income from official sources. In order to be a British Army officer, it was essential for a man to have private means, and most either enjoyed an allowance from their families or were able to use inherited wealth to maintain their position in society. Any young man wishing to become an officer, but lacking the necessary financial endowments, could join the Indian forces. In India the cost of living was low. Servants were cheap and plentiful. The officers of the local forces were not expected to keep up the same level of expenditure as those of the British Army, nor did British Indian society offer the same opportunities for conspicuous spending as that in the United Kingdom.

The average officer of the Indian service was able to enjoy a far higher standard of living than that to which he could have aspired at home. Most, indeed, through their comparative indigence, could not have entered the British Army at all. Few could have employed numerous servants, or maintained numbers of horses, or taken part in expensive field sports. Nevertheless, there were disadvantages to prolonged Indian service. These included voluntary exile from family and friends, exposure to a bad climate and tropical diseases, and life amid strange, sometimes hostile, peoples. There were, however, pecuniary compensations. Officers of the British Army serving in India received extra allowances, and officers of the local forces permanently stationed there received further allowances, which had the effect of rewarding them for full-time expatriate service.

Service in India was not popular with the British Army until late in the nineteenth century. Many senior British officers (Lord Cardigan was an extreme example) despised those who had been to India, and considered nothing of social or military value was to be gained by service there. In most cases appointments in India were accepted because of the financial inducements offered. A British general appointed to one of the three commander-in-chief posts in India, or to command of a division in the Indian Army, was in most cases going to an employment with better prospects than he could find elsewhere. Officers given other Indian staff appointments were well paid for their services, and those who went to India as regimental officers could expect to benefit by doing so.

Commonly, those who went from the British Army did so because by going they gained promotion in the place of a man who had decided not to go, or because they could afford to transfer from the infantry to the cavalry (always an expensive arm) in India, but not in the United Kingdom. Beau Brummel resigned his commission in the Hussars rather than move with the regiment when it was transferred from the fashionable resort of Brighton to the cotton-mills of Manchester in the barbarous north of England. How much more reluctant were officers to leave England altogether and sail with their regiments to India. The number of officers who resigned or transferred rather than 'go East' with their men was rather a joke in the Army, but it was considered in no way discreditable or even particularly remarkable that they should do so.

The facility with which officers could decide whether or not to serve in India resulted from the system by which officers in the British regular army actually purchased their appointments. The Crown, through the Commander-in-Chief of the British Army, granted commissions to suitably qualified men when a regiment was raised. In return for his commission, which carried the right to command men and to receive the appropriate pay of the rank for as long as he held it, the officer paid a fixed sum which was used, notionally, to cover the cost of raising and equipping the regiment. The commission was therefore virtually the officer's own property. When he left the Army, he sold it to his successor and thus recouped his initial outlay. The successor in turn sold it to his successor and so on. Anyone with the money to buy a commission and able to fulfil (or evade) certain minimal physical and educational standards could thus become an officer.

The highest rank open to purchase was that of the colonel of a regiment. Promotion to major-general and above was by seniority only, and staff or extra-regimental appointments were filled by selection. When a senior officer (for example, a major) left the regiment, he offered his commission for sale to the officer next to him in rank in the regiment, in this case the senior captain. If this officer could not afford to find the difference between the cost of a major's commission and his own captain's commission, which he would offer to the senior lieutenant, then any other captain in the regiment might buy it. If no regimental captain could buy it, it would be offered to captains outside. A rich lieutenant could not buy his majority direct, but could buy a captaincy outside the regiment if none existed in his own, and then rejoin immediately, by buying, as a captain, the vacant majority. Thus an officer had, on each promotion, to find only the difference in price between his old commission, which he sold, and his new one. Exchanges between officers of equal rank, either from one regiment to another, or from active employment to half-pay, were equally common.

Long strings of exchanges and purchases might be involved when an appointment fell vacant, in much the same way that long chains of contracts are often involved when a house is bought or sold. Official tariffs were authorised as shown in Table 7. A lieutenant's commission cost much the same as a small house in present-day monetary terms, a colonel's commission was worth about the same as an estate, or manor house and park. Firms of Army Agents were set up for the convenience of officers, buying and selling commissions and arranging exchanges in the same way as estate agents deal with houses and property. They could also arrange a loan for an officer without capital, so that he could buy his promotion, paying the interest out of his pay, in the same way that brokers arrange mortgages or house loans for the would-be owner-occupier.

There were advantages to the State in this system. The officer who left, or 'sold-out', realised a large sum in cash, and was therefore clearly in no need of a pension. The prospect of realising this sum, and of doing so at any time convenient to himself, made it easy and attractive for an officer to leave the Army, so encouraging promotion and avoiding stagnation in the senior ranks. From a more serious point of view, the money invested in a commission was a form of bond for good behaviour. A cashiered officer had his commission revoked and could not sell it. Curiously, until just before the Crimean War, an officer who

Table 7

Prices of commissions in the British Army, 1855

Rank	*Full price of commissions*	*Difference in value between the several commissions in succession*
	£	£
Life Guards		
Lieutenant-colonel	7,250	1,900
Major	5,350	1,850
Captain	3,500	1,715
Lieutenant	1,785	525
Cornet	1,260	
Royal Regiment of Horse Guards		
Lieutenant-colonel	7,250	1,900
Major	5,350	1,850
Captain	3,500	1,900
Lieutenant	1,600	400
Cornet	1,200	
Dragoon Guards and Dragoons		
Lieutenant-colonel	6,175	1,600
Major	4,575	1,350
Captain	3,225	2,035
Lieutenant	1,190	350
Cornet	840	
Foot Guards		
Lieutenant-colonel	9,000	700
Major, with rank of colonel	8,300	3,500
Captain, with rank of lieut-colonel	4,800	2,750
Lieutenant, with rank of captain	2,030	850
Ensign, with rank of lieutenant	1,200	
Regiments of the Line		
Lieutenant-colonel	4,500	1,300
Major	3,200	1,400
Captain	1,800	1,100
Lieutenant	700	250
Ensign	450	
Fusilier Regiments and Rifle Corps		
First lieutenant	700	200
Second lieutenant	500	

NB These prices were generally increased by unofficial extra payments. These 'over-regulation' payments, which were openly recognised by the authorities, could add up to 50% to the cost of commissions.

died, whether from enemy action or other cause, also lost the value of his commission. His widow and estate did not benefit, but it was given, free, to his junior officer, who thus obtained promotion and a commission he could sell for its usual price when the time came. Hence there was a popular toast among junior officers, to 'a bloody war or a sickly season'. It is to the great credit of these officers, who were thus asked to hazard their property as well as their life and limb on active service, that they behaved throughout with as much courage and devotion to duty as officers of other armies.

In the early days of the standing army, public opinion in the United Kingdom had still reason to fear a military *coup*. The insistence on the purchasing of commissions according to a proper scale meant that military power was entrusted only to men with a stake in the country, and the greater power they held the greater stake they had to have. It was held that while poor men, if officers, would have a temptation to better their lot by supporting a revolution, rich men, who had much to lose, would be unpromising material for agitators.

There was, it is true, the consequential disadvantage that men of ability but without means could not expect to serve as officers, at least not beyond the junior ranks. An officer unable to afford to buy his 'step' when the time came had to remain where he was, no matter what his merit and experience. Frequently he had to watch richer men, of lesser ability or service, buy promotion over his head, and a major or captain might easily be ten years younger than his senior lieutenants. Once again it may be remarked that it was greatly to the credit of such officers that they continued to do their duties honourably despite the apparent unfairness which they suffered. However, everyone knew the way the system worked when he joined, and from that point of view could have no complaint. One compensating factor was that when a regiment was sent to a difficult or unpleasant station, such as India, the officers who accompanied it contained a high proportion of seasoned campaigners, who were serious full-time soldiers, looking to the Army for their livelihood rather than a pastime, and who in such circumstances were the sort of officers most needed.

The *Army Lists* tell their own tale of the exodus of officers from British regiments warned for active service. The names of officers in a regiment before it is warned for India are by no means the same as those listed after the regiment has arrived there. In some cases the number of changes is small, but in others it is high, up to 50 per cent or more in each rank. The evidence of the *Lists* must be treated with caution. Every unit has a wastage or turnover of its personnel, wherever it is stationed. Some of the changes would have occurred irrespective of the Indian posting, as officers retired, died, were promoted, or exchanged into other regiments for personal reasons. Some of the officers listed with their regiments in India would be in fact at home, at the depot or recruiting, or in some staff or other extra-regimental employment. Nevertheless there is a higher incidence of these 'casualties' (as the Army termed such casual occurrences) in regiments sent to India than in those remaining in the United Kingdom, and many of the names of officers exchanging into regiments warned for India are in fact those of officers in regiments already there and due to return. These were men for whom there was some incentive to remain in India. The very fact that so many officers left regiments warned for India meant promotion or advance of seniority for any who cared to remain.

The officers of the Royal Artillery and Royal Engineers did not purchase their commissions. In the case of these arms (which until 1855 were administered by the Master-General of the Ordnance and were under the Commander-in-Chief only for operational purposes) it was realised from an early date that some professional and technical training was desirable. In 1741 the Royal Military Academy was founded at the Depot of the Royal Artillery at Woolwich. Thereafter, all aspirants for commissions in the 'ordnance corps' had to attend as gentlemen cadets at the RMA and successfully complete the course of instruction in military and academic subjects. Those passing out highest in order of merit had the first choice of the limited number of commissions in the Royal Engineers, which thus, until 1939, had the intellectual cream of the British Army at its disposal. Officers in these ordnance corps obtained their subsequent promotion by seniority.

In 1799, Colonel, later Major-General, John Gaspard Le Marchant, a cavalry officer who had studied his profession seriously and a man of drive and high intellectual power, secured royal support to found the Royal Military College. Le Marchant had been among many disturbed by the poor performance of the British Army in its early campaigns against the French revolutionaries. While British officers could be as brave, and, given time, as competent as the officers of other armies, it seemed wrong that the time for them to learn their duties should be bought on the battlefield, at the expense of their country's interests and their soldiers' lives. Therefore the Royal Military College, which moved into its newly built quarters at Sandhurst in 1812, was set up to train gentlemen cadets for the cavalry and infantry, just as Woolwich trained them for the Ordnance.

The founding of the RMC did not affect the purchase system. A gentleman cadet successfully completing the course would be granted a free commission as cornet or ensign which he could sell on promotion in the usual way. However, this financial incentive was not sufficient, in the society of the time, to make it particularly attractive to go to Sandhurst, and most young men preferred to enter the Army directly as subalterns by purchase. There were even cases of gentlemen cadets who failed either the Woolwich or the Sandhurst course and subsequently bought commissions in the Line. Le Marchant's original plan of offering a number of free places at the Royal Military College to the sons of deserving officers who had died or were in reduced circumstances was abolished in the 1830s as it was officially considered undesirable to introduce, into the peacetime Army, officers without means of their own. The major advantage of going to Sandhurst was that its cadets were guaranteed a place in the Army, since would-be applicants for entry by purchase could not be sure that they could find a vacancy, and at times there were up to twenty applicants for a single post.

Not until 1870, when the startling rapidity of the success of German arms during the Franco-Prussian War led British public opinion to insist on British officers being properly trained, was the purchase system abolished. The British officer could no longer buy and sell his commission, although he did receive, in compensation for this, a non-contributory superannuation benefit in the shape of a retirement pension which could be commuted into a lump sum. For a time, commissions for cavalry and infantry were granted by open competitive examination. As there were more successful candidates than there were vacancies for them to fill, it was decided to send them to the Royal Military College for a year's instruction. The system was not greatly successful. Young officers of eighteen found that the rules drawn up for cadets of fifteen were still in force. They were not allowed, for instance, to keep dogs in the College, although an officer's dog was, indeed still is, almost as much a symbol of his status as his sword or badges of rank. They were forbidden to smoke, since tobacco was regarded by the authorities in 1870 with the same horror as cannabis in 1970. In 1877 the former system was revived, except that the gentleman cadet was now of the average age of eighteen. A number of free cadetships, held by so-called 'Queen's Cadets' were offered to the sons of deserving officers, and the sons of presumably less deserving officers were charged fees at a lower rate than the sons of civilians, thus restoring the system by which the College was financed prior to 1870.

British officers and the Indian service

Attitudes changed in the latter part of the century, so that it was no longer considered honourable to leave one's regiment when it was sent to India. The coming of the steamship and the opening of the Suez Canal made the country seem less far away. Improved standards of hygiene brought down the mortality rate. Exchanges still continued, if they could be justified on medical or personal grounds, but were officially frowned on. But the lower cost of living in India continued to attract the less well-heeled British officers to regiments there. In 1907 the future Field-Marshal Montgomery, having failed to obtain a place in the Indian Army, chose a British regiment, the Warwickshire Regiment, mainly because it was stationed in India where he could afford to be an officer, rather than in the United Kingdom, where, at that time at least, he could not. In 1896 the young Winston Churchill, as an impecunious lieutenant of the 4th Hussars, was extremely reluctant to go to India, looking on the prospect of service there as 'utterly unattractive, . . . useless and unprofitable exile'. Once there, however, and sharing nearly thirty servants with two other subalterns, he had to admit that whatever the country lacked in suitable company it made up for in comfort. Moreover, he had deliberately chosen to join the 4th Hussars partly because it was about to leave for India, and he expected the consequent augmentation of its numbers to give him more rapid promotion, as there would be several officers joining under him.

But however poor a British Army officer might be, and however much time he spent in India on that account, he was still quite clear that he was of superior clay to any Indian Army officer. The young Churchill, grandson of a duke and a member of a fashionable cavalry regiment, had little time for any of the British community in India, the Anglo-Indians as they were at that time called. Apart from the daughter of the Resident at Hyderabad, with whom he fell in love, he looked down upon Anglo-Indians in general, describing them as supremely uninteresting people, and their women as nasty vulgar creatures.

Bernard Montgomery, with fewer social pretensions, was equally critical. Writing in his *Memoirs* he noted that although the soldiers of the Indian Army were splendid military material and natural fighting men, not all the officers were so good. The climate and prolonged absence from Europe sapped their vitality, and they tended, in his opinion, to age rapidly after about forty-five. He also noted that many were described as 'good mixers', which meant to him good mixers of drinks, as he noticed that the term good mixer seemed to be applied to a man who never refused a drink. He felt grateful, after all, that he had not been able to gain an Indian Army vacancy.

It would be wrong to make too much of the letters of one home-sick young man or of the memoirs of a field-marshal whose puritanical attitude to the consumption of alcohol and tobacco was a by-word throughout the Army, but they are fairly typical of the approach of many British officers to their Indian counterparts. Indian Army subalterns, who under the Crown had to spend a probationary year with the British Army, were sometimes made to feel their inferior position very keenly. It was said, indeed, that one British regiment actually made them dine at a separate table in the Mess!

Officers of the Indian service

In the early days of the East India Company's army, the officers of the local service were a rather mixed body of men. The Court of Directors having resolved to employ no gentlemen as its servants, the officers of its army were necessarily drawn from other stations in life. Pay was extremely low, and attracted none with pretensions to gentility, or indeed even to respectability. In 1753 it was stated that one of the Company's military officers had been a trumpeter at a travelling circus in England, while another had previously been a barber. The Governor of Bombay's steward became lieutenant of the grenadier company, and carried on his military and household duties simultaneously. Lieutenant Stirling, appointed in 1740 to the vacant captaincy of the Bombay garrison, was accepted despite the disadvantage of being unable to read or write, which, his Commander-in-Chief pointed out, 'might be inconvenient in an emergency'. Accordingly Lieutenant Thomas Andrews was made his assistant and proved especially useful in carrying on the local rice trade, a monopoly held by this office.

In the 1750s and 1760s efforts were made to recruit a better class of men. The expansion of the Company's army increased the number of posts available, at the same time as reductions occurred in the British Army (at the end of the War of the Austrian Succession and the Seven Years War). A number of officers who had served in these wars, in Europe or elsewhere, and who did not wish to go on the half-pay or unemployed lists, were able to continue in full-time military employment by transferring to the Company's army.

In 1794 it was decided that the Company's officers should be granted commissions in the name of the Crown, by the Commander-in-Chief, India. These commissions were local, and temporary, to enable them to command troops of the British Army in India. West of the Cape of Good Hope, they ceased to have any effect. No officer of the East India Company, no matter how much he had achieved in the service of the British Empire in India, could appear at his sovereign's court dressed in uniform. Instead he had to appear in the civilian clothes of a private gentleman. Even after 1858, when the Company's officers were transferred to the service of the Crown, officers of the Indian forces held a lower status than those of the British Army, and British officers took precedence over those of equal rank in the Indian service. There is no doubt that officers of the local Indian service were drawn, on the whole, from classes lower in the social or financial scale than those of the British Army.

Some 3,500 officers served in the Company's regiments of the Bengal Army at some time between 1820 and 1834. In about 2,000 cases, the occupation of their fathers is given in V. C. P. Hodson's biographical lists *Officers of the Bengal Army*. An analysis of the information provided from these (see Table 8) shows that they were predominantly drawn from the ranks of the British middle class, and some were from working-class families. A few were from titled families, but these were either younger sons who had to make their own way in the world, or from the nobility or landed gentry of Ireland, which was much less affluent than that of England. Out of 2,000 officers, one was the son of a marquis, four were sons of earls, one the son of a viscount, six sons of barons, and sixty-six sons of baronets. Four were illegitimate sons of earls and one of a baron. Only one officer succeeded to a peerage, the second son of the Earl of Carnwath. Six officers inherited baronetcies, but only three were eldest sons. On the other hand, a large number of officers' mothers, and rather fewer of their wives, were the daughters of titled families.

Table 8

Class origin of Bengal Army officers, 1820–1834

(showing the occupation of the fathers of 1,945 officers who served with the Bengal Army at some time between 1820 and 1834)

I *Occupation of officers' fathers who were not employed in India*	*Number in that occupation*	
Army officers	331	
Navy officers	86	
Marines	2	
Total in the services		419
In clerical orders		307
Merchants		118
Barristers	47	
Solicitors	46	
Judges	3	
Total in the legal profession		96
Tradesmen		68
Doctors	51	
Surgeons	24	
Total in the medical profession		75
Bankers		37
Customs officials		28
Farmers		27
MPs		17
Clerical workers		22
Craftsmen		14
Manual workers		18
Officers of East Indiamen		7
Worked outside the UK but not in India		30
Others		120
	TOTAL	1,403

Table 8 (*cont.*)

II *Occupation of officers' fathers who worked in India*		*Number in that occupation*
Officers in the Indian Army		252
Civil servants not included separately		96
Surgeons	34	
Doctors	4	
Total in the medical profession		38
Merchants		32
East India Company Naval Service officers		31
Judges		27
Attornies		12
Commercial residents (businessmen)		11
Indigo planters		7
Clerks for commercial firms		4
Chaplains		3
Bankers		1
Others		28
	TOTAL	542

Many of the officers were from military families; 331 had fathers who had been officers in the British Army, of whom a number were likely to have served in India; 86 officers were the sons of Royal Naval officers, of whom 26 were admirals.

After the armed services, the largest occupation represented was the Church. The 307 clerics whose sons were in the Company's service during this period included one bishop, several deans and archdeacons, and a number of university teachers. The majority were ordinary parish priests, and such men, although for the most part comfortably off, would have lacked the means to buy a commission in the British Army if they had several sons to provide for.

A high proportion of officers came from families in government administrative posts. Twenty-eight fathers held offices, at different levels, in the Revenue department, ranging from the Assistant Comptroller of Customs for Scotland to excise officials in small towns. One of the latter was the poet Robert Burns whose third son joined the Bengal Army. Other officials included the principal clerk at the War Office, the Deputy Treasurer of Greenwich Hospital, the secretary to the Receiver-General for Wiltshire, a Clerk to the House of Commons and a Comptroller of the Lottery. Some posts were clearly ones held by gentlemen of high position, such as the Garter King of Arms, a Gentleman Usher to George III, and the Lord Chancellor of Ireland. Others were lower in the social order, such as the eight writers to the signet, or the two town clerks. The East India Company's minor officials, and those of the Board of Control, were able to secure, between them, twenty-seven nominations for their sons.

Of the non-official classes, the legal profession was represented by ninety-six entries, and the medical by seventy-five. There were thirty-seven sons of bankers and seventeen of

Barracks and parade ground, Dagshai, Punjab. Note the bullock carts in the foreground. Photograph 1904

(above) *Royal Artillery Mountain Battery, 1880 (British gunners and Indian drivers with mules)*; (below) *'Saving the Guns at Maiwand', July 1880. E–B Battery RHA coming out of action. Photogravure after G. D. Giles*

Members of Parliament. Such occupations, while respectable, were rarely considered ones suitable for noblemen or persons of aristocratic birth, and many British commanding officers would not have considered persons of such descent as suitable to hold the King's commission, even if they could afford so to do. (Indeed, one commanding officer of a Household Cavalry regiment referred to his troopers as 'damned cheesemongers' when entry to the ranks was opened to the sons of wealthy City merchants.)

A total of 118 fathers of Bengal officers are described as merchants, and in addition a number of tradesmen appear in the lists. These included three hairdressers, two hatters, a hosier of Nottingham who secured places in the Bengal Army for two of his sons, a grocer, a saddler, an upholsterer, four drapers and four booksellers. There were also twenty wine merchants, five dealers in spirituous liquors, an innkeeper, two hop merchants, three distillers, and eight brewers.

Craftsmen whose sons became Bengal officers during this time included two silversmiths and a goldsmith, three cabinet-makers, two stone-masons and a builder's assistant.

A number of fathers were connected with the lower echelons of naval or maritime affairs. Such included two master mariners, two ship's masters, and two pursers from East Indiamen. Such men held important technical appointments but were not commissioned officers, and were definitely lower in the social hierarchy than naval officers.

A further 542 fathers whose professions are listed were connected with India in some way. Nearly half of these were officers of the Company's army, and a strong tradition of employment in India was maintained by many families. Many officers had uncles, cousins, or grandfathers who served the Company in India. Sometimes whole families served together. The four Bellew brothers all entered the Bengal service; the eldest died in India aged eighteen, the second retired after earning his pension and died at the age of sixty-nine, the third died in India at the age of twenty-five, and the youngest was killed in action in the First Afghan War aged thirty-nine. Among the 252 officers whose fathers had been in the Bengal Army, there were eighteen sets of brothers. Cousins often served together, such as the three who were grandsons of the actress Sarah Siddons.

A complicated, but not untypical example of the close-knit relationship of officers at this time is provided by the Skardon family. Samuel Skardon (1730–88) was a lieutenant-fireworker and commissary of ordnance who died in India. Of his children, one son became an officer in the Bengal Native Infantry and eventually rose to the rank of lieutenant-general, two daughters married officers of the Company's service, and a third daughter had a son who joined it in due course. Samuel's widow re-married twice, and had another son who also joined the Bengal Army.

Thirty-one officers of the East India Company's naval service (its warships as distinct from the chartered East Indiamen) sent their sons to the Bengal Army, as did 123 members of the covenanted Indian civil service.

An important result of the officers' being drawn predominantly from the less affluent sections of the British middle class was that junior officers were unable to meet the expenses of their appointments. Uniform and field equipment cost about Rs 2,000. Few parents or guardians of East India Cadets could afford this outlay. The young subaltern on being commissioned had to borrow the money, either from firms of army agents at an interest of about 10 per cent per annum, or from Indian moneylenders who charged about 24 per cent. Life insurance, demanded by the army agents before they would advance money, added another 3 or 4 per cent to their terms. A subaltern could scarcely pay the interest on such loans, far less the capital. As his original equipment wore out and had to be replaced, further debts were incurred. An officer could not begin to clear his debts until he reached the rank and pay of a major, which is one reason why so many officers stayed in the Army in order to reach that rank.

The living expenses of a lieutenant in Bengal in 1831 were about Rs 275 per month. This included such items as Rs 8 to the pension funds (for officers' widows and orphans, etc), Rs 40 for house rent, Rs 45 for meals, Rs 35 for wines, Rs 6 for candles, Rs 59 for servants, Rs 6 for cigars, and Rs 17 for horse-feed. The result of this expenditure, on a pay which even with full allowances was only Rs 365 per month, meant that, on average, subalterns of the Company's armies were at this time in debt to the tune of some Rs 7,000 (or £700).

The rates of pay of an East India Company's officer in India, and of a British Army officer in the United Kingdom, in the middle of the nineteenth century, are given in Tables 9 and 10.

Unlike the officers of the British Army, the East India Company's officers did not buy their commissions. Indeed there was some doubt as to whether they held real commissions at all, since although the Company's charters gave it the right to grant commissions, their wording suggests that these commissions were to individual appointments rather than general warrants to exercise and command troops, and the relevant statutes never refer to commissioned officers in these express terms, but rather to military officers, or persons serving as officers. They were, at the very end of the eighteenth century, granted temporary and local commissions in the name of the Crown by the Commander-in-Chief, India, who had the right to do this by the terms of his own appointment. The

Table 9

Pay of East India Company's officers plus maximum allowances, 1855

The official figure is that given in rupees *per month*. The figure given in sterling is the equivalent *daily* rate, for ease of comparison with the British Army pay. The exchange rate of the rupee varied. At the end of the nineteenth century it was stabilised at 15 to the pound, equal to one shilling and four pence. In 1855 it was about 10 to the pound, equal to two shillings, and this figure has been used in the calculations below.

	Horse Artillery and Cavalry	*Foot Artillery and Engineers*	*Infantry*
Colonels	Rs 1,478	Rs 1,295	Rs 1,295
	£4 16s 7d	£4 5s od	£4 5s od
Lieut-colonels	Rs 1,157	Rs 1,032	Rs 1,032
	£3 16s od	£3 7s 11d	£3 7s 11d
Majors	Rs 929	Rs 789	Rs 789
	£3 1s od	£2 11s 11d	£2 11s 11d
Captains	Rs 563	Rs 434	Rs 415
	£1 11s 5d	£1 8s 5d	£1 7s 2d
Lieutenants	Rs 365	Rs 266	Rs 257
	£1 4s od	£0 17s 5d	£0 16s 10d
Cornets, 2nd lieuts	Rs 311	Rs 213	Rs 203
and ensigns	£1 os 5d	£0 14s od	£0 13s 2d

Table 10

Daily pay of officers of the British Army in the United Kingdom, 1855

	Life Guards and Horse Guards	*Foot Guards*	*Dragoon Guards and Dragoons*	*Foot*	*Royal Artillery*		*Royal Engineers*	*Royal Marines*
					Horse Brigade	*Foot*		
	£ s d	£ s d	£ s d	£ s d	£ s d	£ s d	£ s d	£ s d
Colonel	—	—	—	—	1 12 4	1 6 3	1 6 3	—
Lieut-colonel	1 9 2	1 6 9	1 3 0	0 17 0	1 7 1	0 18 1	0 18 1	0 17 0
Major	1 4 5	1 3 0	0 19 3	0 16 0	—	—	—	0 16 0
Captain	0 15 1	0 15 6	0 14 7	0 11 7	0 16 1	0 11 1	0 11 1	0 10 6
Lieutenant	0 10 4	0 7 4	0 9 0	0 6 6	0 9 10	0 6 10	0 6 10	0 6 6
Cornet, ensign, and 2nd lieut	0 8 0	0 5 6	0 8 0	0 5 3	—	0 5 7	0 5 7	0 5 3

NB At this time there were no majors in the Royal Artillery and Royal Engineers.

Governor-General, whether under the Company or the Crown, had no powers to grant commissions. The so-called Viceroy's Commission held by Indian soldiers serving as officers with the ranks of subedar and jemadar thus had no constitutional validity as a military commission.

The usual method by which a young man obtained an appointment as an officer in the East India Company's army was to obtain a nomination from a Director of the Company. After securing his nomination the candidate applied formally to East India House with documentary evidence as to his date of birth or baptism, health, education, and good character. He then appeared before the appropriate committee of the Court, which, if satisfied by his application, appointed him as a cadet and arranged for his departure to India or to a period of training in the United Kingdom.

Officer training in the Company's army

The need for training applied especially to candidates for the artillery and engineers, for it was apparent that whereas any young gentleman who could stay on a horse, or could put one foot in front of the other, was suitable to serve with the cavalry or infantry respectively, some further education was required of those serving with the technical arms. There was constant difficulty in filling such appointments in the early days, and recourse was had to the gunners and gun-room crews of East Indiamen, to foreign mercenaries, to officers transferred from the infantry, or to officers of the royal service wishing to join the Company's army.

Not until 1798 was it determined that future officers for the Company's ordnance should have professional training. An additional ten cadetships were created at the Royal Military Academy to be held by East India Cadets. Subsequently the Company was allowed to fill forty of the hundred cadetships at Woolwich, but East India Cadets were not allowed to compete for vacancies in the Royal Army, and vice versa. The RMA was unable to find space for all those wishing to attend, and so in 1803 an ordnance company was formed at the Royal Military College, in which twenty places were reserved for East India Cadets. The Directors paid £100 each year to the Academy or College for the tuition, board, and equipment of each cadet.

The arrangement was not entirely satisfactory, and with the expansion of the British Army during the Napoleonic Wars there was an increasing requirement by the Royal Corps for available officers. The Court of Directors therefore decided to form its own officer-training establishment, the Military Seminary, which opened in 1809 in the converted mansion of Addiscombe House, near Croydon on the southern outskirts of London. Because the requirement was primarily for ordnance officers, the cadets wore the blue uniform of the Company's artillery in the same way that those at Woolwich wore that of the Royal Artillery. The organisation and syllabus of the Royal Military Academy were closely followed.

Cadets were taken in at the average age of fifteen and commissioned after satisfactorily completing a two-year course. Military officers provided the command and staff, and shared with civilian professors the duties of instructing in such subjects as fortification, mathematics, landscape and military drawing, chemistry, geology, French, and Hindustani. As well as these academic subjects the cadets were taught drill, both gun and musketry, and the use of the sword. Posts, academic and domestic, at the Military Seminary were filled by the patronage of the Directors. Cadets were required to pay £30 per annum (later £40) for their education, the rest of the £75 pa which was the actual cost of maintaining each cadet being borne by the Company. The civilian professors, who were almost entirely men with high academic qualifications, had their salaries governed by age and experience. A junior man might be paid only £80 or £100 pa, a senior £500 or £600. They were also granted substantial sums for their board and lodging, and many supplemented their income further by the publication of learned works on their subjects. John Shakespeare, for instance, Professor of Hindustani from 1809 to 1829, received £3,600 from the Directors alone for his Hindustani grammar and dictionaries. Jonathan Cape, Professor of Mathematics, received an additional £50 for filling the office of chaplain; Samuel Parlour, Professor of Classics, who was popularly believed, from the high stock which he always wore, to have tried to cut his throat when it was pointed out to him by Cape that the calculations by which he claimed to have squared the circle were wrong, obtained an extra salary of £50 by acting as store-keeper.

As Addiscombe could house more than the number of ordnance cadets required each year it was decided that the Seminary should be filled up, and that the final order of merit should decide whether the cadets were appointed to the engineers, artillery or infantry. In the first twelve years of its existence the Military Seminary passed 62 cadets to the engineers, 215 to the artillery, and 113 to the infantry. Young men wishing to join the infantry as cadets could do so by going directly to India, and no cavalry cadets at all went to Addiscombe.

There was between 1804 and 1811 a college at Barasat in Bengal, where these cadets were to be taught Hindustani, drill, and minor tactics. However, the only military subjects at which the cadets shone were the traditional ones of drinking, swearing and duelling. Officers were attacked, the police went in fear, and the local natives were oppressed and abused, so that the establishment was disbanded after seven years.

Addiscombe, too, had its disagreements with the locals, although these were more in the tradition of Woolwich and Sandhurst than Barasat. Intermittent 'town versus gown' riots took place throughout the history of all three British establishments. Addiscombe's greatest moment was at the time of the Chartist disorders of 1848, when a supply of ammunition, for their muskets and 3-pounder guns, was sent by the government authorities to the Seminary in case it became necessary to use the cadet company in quelling civil disturbances. The cadets eagerly looked forward to the bombardment of Croydon, and to driving its citizens from the smoking ruins at the point of the bayonet. However, no such necessity arose, and the cadets had to learn the duties in aid of the civil power after they arrived in India.

The promotion system in the Company's army

Once in India, the newly appointed officer or cadet hastened to join his regiment, since although he received a free passage on the Indiaman, his pay and seniority did not begin until he actually joined. Thereafter promotion went by regimental seniority in strict succession, and nothing could deprive an officer of promotion except a court-martial. This method, although fairer than the purchase system, made promotion extremely slow (except in time of war or pestilence) and allowed men who were unfitted by age or physical infirmity or even military incompetence to attain senior rank. In order to encourage officers to leave, a system grew up in some ways similar to that of purchase, known as the 'subscription system'. An officer who wished to leave would be paid to go by his juniors, who thus all moved one place nearer to promotion. A substantial lump sum was thus given to the retiring officer to compensate him for the higher pay and pension he would have received had he stayed on to await his turn for promotion. Proper rates were drawn up, and each officer subscribed in accordance with the amount laid down for his rank. The amount varied in accordance with the immediate improvement in the prospects of subscribers. Thus, out of the Rs 30,000 (£3,000) which was the total subscription given to an infantry major, Rs 12,000 would be subscribed by the senior captain and only Rs 150 by the junior ensign.

Some idea of the effect on the average age of senior officers of promotion by seniority may be gained from Table 11. Comparable figures for the British Army are shown in Table 12. Royal Engineers and Royal Artillery figures are not given, since at this time they did not serve in India, nor did they have regimental majors.

While it would be wrong to draw too sweeping a conclusion from the figures for one year only, selected at random, they are of interest, since they suggest that the age of promotion for British officers in the army as a whole was significantly lower than that of

Table 11

Average age and service of majors in the East India Company's armies, 1855

	Average number of years' service before promotion to major	*Average number of years' service in that rank, in 1855*	*Average age of serving majors*
Light Cavalry			
Bengal	30	3½	52
Madras	23½	7½	48
Bombay	27½	3	48
Artillery			
Bengal	30	3½	52
Madras	27	5	50
Bombay	32	2	52
Engineers			
Bengal	27	5	50
Madras	29	1	48
Bombay	26	1	45
Infantry			
Bengal	27	3	48
Madras	27	4	49
Bombay	27	3½	48

Colonels were often in their sixties and generals in their seventies.

Table 12

Average age and service of majors in the British Army, 1855

	Average number of years' service before promotion to major	*Average number of years' service in that rank, in 1855*	*Average age of serving majors*
Cavalry			
British Army as a whole	14	2	33
British Army in India	17	3	37
Regiments of Foot			
British Army as a whole	11	4	32
British Army in India	21	4	42

those serving in India. It is true that in 1855 the British Army had expanded as a result of the Crimean War. But, although this expansion allowed men to be promoted at an earlier age than would otherwise have been possible, the battlefield casualties allowed many senior men, who had been unable to afford to buy their 'step', to be promoted without purchase, thus bringing to the higher ranks an influx of men with more years and service. The British regiments in India, although missing the battles of the Crimea, had in many cases seen action in the Punjab or Burma, and could be considered on active service even in peacetime, at least as far as the ravages of the climate and disease were concerned. The greater age of their senior officers does support the argument that many of the British Army officers who served with the garrison of India were men less affluent and therefore less able to buy promotion than those who remained with the Home Army.

Extra-regimental employment

An officer who stayed with his regiment had little to look forward to for the first thirty years of his service, the years when he would be at his most active and enterprising, except the routine duties of a troop or company officer. All those who could do so obtained transfers to newly raised irregular regiments, in which there were only a few officers all holding interesting and well-paid appointments, or to civil or political employment, where the jobs were even more interesting and even more well-paid. Staff appointments, which carried good emoluments and higher prestige than regimental duties, were eagerly sought after and reluctantly given up.

Sir Charles Colville, a British general, commander-in-chief of the Bombay Army in the early 1820s, expressed to the Governor, Mountstuart Elphinstone, his dissatisfaction with the arrangement whereby officers had to leave staff appointments on obtaining promotion in the Army. This was because there were so many graded to be held by majors and so few by lieutenant-colonels.

> The Army will still want the active and efficient Battalion Commanding Officers, which the former are generally more suited to than the latter, who from increased years and desuetude of military or other active habits, with the accompanying draw-back of diminution of income, will go (for in some instances it cannot be called returning) reluctantly and of course unprofitably to Regimental duty.

Officers clung to staff appointments for years, stagnating the work of their departments and blocking the opportunities of younger and keener men to enjoy the experience and emoluments of these posts. Accordingly a detailed table of political, civil, and staff appointments was maintained, which could not be held by an officer of a higher rank than the one specified. In this way, when an officer was promoted, as in due course he had to be by virtue of his seniority, he could be removed from his post without acrimony or any outward suggestion of inefficiency. Good men could be retained in staff or civil employment but transferred on promotion to an appointment of higher grade.

This subject continued to be a problem. Writing to a member of the Council of India in 1874, Major-General Sir Henry Green, a former commandant of the Sind Frontier Force, spoke of the unfortunate consequences of mixing military and political appointments. The frontier, he said, had become

> a nursery for squalling children and an asylum for worn out men, some deaf and blind, others incapable, from extreme nervousness and physical debility, from mounting a horse, and all but the very juniors looking upon the active duties of a frontier soldier to be avoided, and those juniors who still possess energy are crushed by the knowledge that no exertions on their part can ensure them from being superseded by individuals who swarm around headquarters ready to grasp at any appointment in which they can obtain higher emoluments, so as to allow them to save money to return to England at the earliest opportunity – all this is painful but too true.

Even allowing for Green's own position as an old soldier regretting a change in policies he had himself supported, there was more than a grain of truth in his argument.

A memorial from officers of the Bengal Army to the Directors in 1797 fairly sums up their attitude towards the prospect of obtaining allowances additional to Army pay.

> In our early youth we had exiled ourselves from our mother country, bade farewell to our friends, not for the mere receipt of the present pay of our prospective ranks, but for obtaining those high emoluments, bazaar money, revenue commissions, double full batta, the strong and only inducements which decided our selection of the East India Company's service.

The changes in the regulations at that time, which threatened to reduce the chance of extra income from these sources, were condemned as 'a breach of faith [which] dooms us to perpetual slavery or turns us grey headed beggars on the world'. The Directors objected to this memorial and criticised the officers subscribing to it, remarking that 'instead of receiving the material benefits and advantages held out to them by the new Regulations with gratitude and respect they presumed to arraign those Regulations in a style of disrespect and intemperance highly unbecoming'.

The prospect of service outside the Army was always attractive. The rapid expansion of British dominion in India during the first half of the nineteenth century, and the increasing number of bureaucratic posts in the second far outstripped the ability of the

Indian Civil Service to fill them. There were therefore many chances for officers to transfer to civil employment, and under the regimental system this fatally weakened the units on whose muster rolls they officially appeared, as they were not replaced. The Staff Corps system introduced after the Indian Mutiny, to form officers into a pool from which either military or civil appointments could be staffed, was almost as bad, at least as far as morale was concerned. Sir Frederick Haines wondered how any system of providing officers could be efficient if it was actually designed to give officers every facility for leaving their own profession to enter another one.

Officers who did not find staff appointments – either at headquarters or with the irregular regiments or in civil employment – had to find other ways of improving their pay. The most lucrative method was to find a post at one of the stations deemed to be an active service establishment, where 'batta' or field service allowance, was payable. Batta was intended to recompense the officer or soldier for the additional costs of carriage, food, etc, incurred on campaign, but it was often difficult to say at what moment a campaign ended and peace began. By the time it had been decided, the recipient of batta had enjoyed for some time a higher standard of living, with enhanced wartime allowances in a peacetime economy, and so resented taking a cut. The gradual extension of British rule over areas of different degrees of pacification led to anomalies, with some stations still being paid full batta and others on half rates. The result was that regiments at full batta stations were fully officered and those at half batta stations had usually only their colonel, who was always paid full batta by virtue of his rank, and the adjutant and the quartermaster, who both received extra allowances for being regimental staff officers. All the other posts on the establishment were held by officers on furlough or on civil or staff employment. The lucrative full batta posts were always kept filled, and when an officer holding one went on furlough or to other duties his place was taken by someone from a half batta station. When a regiment from a full batta station was relieved by one from a half batta station, the bulk of its officers transferred to the incoming unit, which in its former locations would have been under strength in the officer ranks.

Indian service under the Crown

After the transfer of the Indian Army from the Company to the Crown, it was decided that all its officers should bear the Queen's Commission. The idea of keeping Addiscombe as a separate college for the Indian service was considered, but it was eventually decided to close it down and train all Indian cadets at the Royal Military College, Sandhurst. At this time Sandhurst trained some, but not all, officers of British cavalry and infantry, for the purchase system was still in force in those arms. Woolwich was expanded to provide extra officers of the royal service who would in future command Indian gunners and sappers, and to take account of the enlargement of the Royal Artillery by its absorption of the Company's European Artillery.

The military patronage of the Court of Directors was vested in the Secretary of State for India. Its civilian patronage had disappeared with the advent of open competitive examination as the means of selecting candidates for office in India. Under regulations introduced in 1862, the Secretary of State had the right to appoint twenty cadets each year to the Royal Military College. These cadetships were open only to the sons of former

Indian servants, and their fees, subject to a means test, were paid out of Indian revenues. The 'Queen's India Cadets' were not obliged to join the Indian service, although in practice few of them were wealthy enough to join the British, and the scheme was certainly intended to fill Indian needs. Sons of former Indian military or civil officers therefore had an advantage over other candidates because, even if they had sufficient means not to qualify for free tuition, there were twenty places reserved for them at the College, and they had only to compete among themselves for these. Unsuccessful candidates could still enter the general entrance competition and attend at their own expense if they passed. A similar number of posts, 'Queen's Cadetships', were offered to the sons of deserving British Army officers, thus reviving Le Marchant's original scheme for the College.

The average requirement of the Government of India in peacetime was for only a limited number of officers. As cadets were selected by regiments according to their order of merit, any cadet wishing to be sure of entering the Indian service had to be among the top thirty of his term. In fact there were sometimes young men in the top thirty who did not wish to join the Indian service, but the nature of cadets is such that few will study when there is no incentive to do so, and although the Government of India may not have had the thirty best cadets in each term, it certainly had the thirty most industrious. Between 1890 and 1922, twelve out of the sixty-six winners of the Sword of Honour (for the best cadet at the RMC) joined the Indian service.

The end of the purchase system in 1870 meant that nearly all aspirants for a regular commission had to attend the Royal Military College (or Academy). Any vacancies left unfilled were open to candidates entering from the militia, or the universities, but in practice the Royal Military College was the usual source from which the Indian Army drew its regular officers until 1947.

In the Company's army the officers had belonged to the regiments as they were organised in 1824 or raised subsequently. They had a right to promotion according to their seniority within their own regiments. This, while fair to men within each regiment, was at times unfair to those outside it, since if one regiment suffered a high number of deaths in battle, or an epidemic, or even an unusually high number of retirements or resignations, those remaining would be promoted more rapidly than officers who had joined the Army at the same or earlier time, but whose regiments had suffered fewer casualties.

The Indian Staff Corps

To remedy this defect, and the disadvantage of having military officers borne on the establishments of regiments but not actually serving with them, the idea of forming a central staff corps, which had been discussed before the Mutiny, was adopted. The officers of each of the three Indian armies were in 1861 grouped into, respectively, the Bengal, Madras, and Bombay Staff Corps. These Staff Corps acted as pools of officer manpower, from which officers were drawn to fill either civil posts, military departmental posts, army headquarters and staff appointments, or regimental appointments. The establishment of British officers with each unit was drastically reduced, and the posts which they held were deemed to be on the staff of the regiments, with the executive command of companies, troops and sub-units theoretically held by the Indian officers. Promotion was now governed by length of service, not seniority. Lieutenants were promoted to captain after eleven years' commissioned service, captains to major after twenty years, majors to lieutenant-colonel after twenty-six years, and lieutenant-colonels to colonel after thirty-one years.

This system had built-in weaknesses. The Indian officers, though nominally in charge of sub-units, had no powers of punishment over minor offenders. The lack of opportunity for advancement led the energetic, younger men who had led the newly raised levies of the Mutiny to take their *jagīrs* and leave. Their places were taken by men promoted by virtue of seniority, so that the situation again came about in which most native officers were promoted after long service in the ranks, at an age too great to act as subalterns, and without sufficient modern education to absorb the additional knowledge required to act as officers. So far from the European officers acting as 'advisers' to the Indian company commanders, the Europeans had to act as the company commanders, advised, on matters relating to their soldiers' domestic matters, by the Indian officers.

Although officers were not permanently allotted to regiments, in practice they tended to stay with them, and once again inequalities tended to arise. Because promotion was now by service, a man might rise to the rank of major, while still on the staff of his regiment as a wing officer, while in other regiments, should the post fall vacant, a captain might undertake the duties of a wing commander. This gave not only additional status, but also extra pay, as pay was composed of the pay of an officer's rank plus the allowance appropriate to the appointment which he held. Thus the monthly pay of rank varied from Rs 225 for lieutenants to Rs 827 for lieutenant-colonels, and that of appointment varied from Rs 150 or 100 for a squadron or double-company officer to Rs 700 or 600 for a cavalry or infantry unit commanding officer.

Officers of the old armies commissioned before or during the Mutiny were given the option of continuing to serve under the old conditions, and were held on local and general lists respectively. Those in Bengal became known as the 'lucky locals', since, as they continued on the cadres of their regiments that had in many cases mutinied and murdered the officers actually present at the time, these survivors soon rose through seniority to be lieutenant-colonels and with the extra allowances payable after twelve years in that rank, equal to more than £1,000 pa, were able to retire on full pension at a comparatively early age.

Officers of the Staff Corps on civil employment received their military promotion at the same intervals as those actually with the Army. They had the right to return to soldiering at any time, and could, having attained colonel's rank, claim promotion to that of a general officer by seniority. At the age of fifty-five officers were considered too old for civil or political duties, their services were placed at the disposal of the Army, and quite commonly they were appointed to important commands. The disasters of the Second Afghan War showed the results of this system which enabled men to hold high commands even though they might have spent their whole career in civilian duties. In 1882 it was ordered that actual regimental commanders and their immediate successors should not continue in office after seven years or after reaching the age of fifty-five, and thereafter the limits should be seven years' tenure or the age of fifty-two. Officers returned to the Army at fifty-five thereafter received promotion, honorary or otherwise, and were immediately retired on the appropriate pension.

This arrangement was long overdue. In the Madras and Bombay armies, where losses in the Mutiny had been light, the 1870s had seen a block in the prospects of obtaining commands, and a degree of seniority in the regimental officers, that damaged the efficiency of the Army. Nevertheless the human reluctance of elderly officers to give up well-paid appointments, and of Army headquarters to force them to do so (for Army headquarters was composed of men in the same circumstances) continued to weaken the Indian Army. An insufficiently ruthless elimination of the unfit allowed many men to hold commands who were no longer physically or mentally capable of their duties. Rather than supersede an enfeebled individual who had once been a good officer, the seniority system allowed him to obtain and keep command, thus risking the reputation and esteem of his unit, and the lives and well-being of his men. To ensure promotion to a senior command, an officer had only to outlive his contemporaries and stay out of trouble.

Even after World War I the problem was still there. The Commander-in-Chief, Lord Rawlinson, wrote in a private letter that his Quartermaster-General was unreliable: 'His Department is in a state of Chaos. We can get no idea of what he has got, what he wants, or what the value of anything is, so it is rather hard to make any decently reliable estimate of our future requirements.' The Department of Military Works was equally inefficient: 'there is an old block head called Rimington at the head of it who wont last long. He is a Major-General'.

The problem of officer employment was present at this time, too. In 1920 Rawlinson had more than 1,200 officers surplus to requirements, mostly men with between four and eight years' service who had been given permanent commissions during World War I because they could not be induced to join by the offer of temporary ones. He hoped that some of the cavalry officers could be found jobs with the tanks and armoured vehicles when India received them, but the rest, having been taken on for service with the Indian troops in the Middle East, were encouraged to join the Arab Legion, or other armies of the emergent Middle Eastern kingdoms under British influence. Even the Constantinople Constabulary was thought of as a possible means of livelihood!

The employment of military officers in civil duties was officially confined to the non-regulation provinces. These were provinces where the disorderly state of the country, or the uncivilised nature of the inhabitants, precluded the immediate introduction of the full legal and administrative system of the regulation or settled provinces. It was in the unsettled areas, where the magistrate might often have to act at the head of armed force in the apprehension of robbers or the suppression of civil strife, that the military officer was a most suitable choice. Many officers, technically soldiers in civil employ, distinguished themselves as administrators and politicians of the highest order, including such men as Sir Henry Lawrence in the Punjab, Sir Arthur Phayre in Burma, and Sir Henry Ramsay in Kumaon.

However, in some cases the use of untrained soldiers to do the work of civil servants was less successful. Lord Ellenborough, Governor-General at the time of the conquest of Sind, despised civilians and loved soldiers, having desired to be one himself. Civil servants with moustaches he called 'Cutcherry [or court-house] Hussars', and he allowed Sir Charles Napier, the military governor of Sind, to get rid of the few civil servants there and place the whole revenue and judicial administration in the hands of officers, who were quite unfitted to deal with complicated questions of land tenure and revenue, and who were in any case obliged to spend much time dealing with local terrorists and obtaining supplies for the occupying troops.

The extensive employment of soldiers in the higher administrative posts was for many years a grievance with the Indian Civil Service. It was wrong that soldiers should be appointed to posts which Parliament had decided should be filled by men gaining their places in open public competition. For example, a cavalry officer with no experience of civil administration was made deputy commissioner in a province where ten out of the eleven deputy commissioners, in charge of districts, were already soldiers. Gradually the military element was reduced, and the promotion prospects of the civil servants improved. In 1876 military officers ceased to be accepted for further posts in all Bengal, the Central and North-West Provinces, and Oudh. In 1885 Sind followed suit, in 1903 the Punjab, and in 1907 Assam. However, Burma continued to offer to soldiers the chance of a civil career, and so did the North-West Frontier Province, which was specially created out of the trans-Indus districts of the Punjab to allow the Government of India to keep the area bordering the Afghan tribal territories in firmer control.

STAFF OFFICERS

The three Staff Corps were amalgamated into one Indian Staff Corps in 1891, and in 1903 the title was abolished altogether as part of Kitchener's reforms. All members of the Indian Staff Corps were redesignated 'Officers of the Indian Army'. The term 'staff' was now required for its more usual meaning, to indicate the body of officers who advise and assist the commander in the preparation and planning of his operations. The Indian Army, like the British Army, was for the first time about to have a properly organised peacetime General Staff.

Prior to this time the Indian Army had followed the British principle of dividing its staff officers into two branches, those of the Adjutant-General, who dealt with matters such as training, discipline, and personnel, and of the Quartermaster-General who dealt with supplies, accommodation, communications, and intelligence. In 1906 the General Branch was set up, under a Chief of the Staff to deal with matters of military policy, army organisation and deployment, mobilisation and war plans, intelligence and the conduct of operations. The 'A' branch continued to deal with all matters relating to the soldier as an individual, and the 'Q' branch with all questions of supply.

Kitchener's reforms gave the Indian Army an organisation which kept up in peacetime the formations in which it was intended to fight in war. There was accordingly a need for a greater number of trained staff officers to man the headquarters of brigades and divisions, and the old system of collecting staff on mobilisation from Army Headquarters or of recalling officers with experience of staff work from other employment was no longer suitable.

To supply the necessary officers an Indian Staff College was set up in 1905 and moved to permanent quarters at Quetta in 1907. Hitherto first appointments in the staff had been largely by the patronage of senior officers, and subsequent employment depended on the personal impression made by officers at headquarters. There was no need for anyone to have attended the Staff College at Camberley, which had evolved from the old Senior Department of the Royal Military College, and therefore few did so. Although since 1886 eight out of the sixty vacancies on each course had been reserved for applicants from India, very few of them were actually taken up. The major reason for this was financial. While he was a student at Camberley, the Government of India reduced an officer's pay from Indian to British rates, on the grounds that an officer studying in his own country could not expect to receive the additional increments which he received by virtue of his expatriate service in India. However, this reduction in the pay of his rank, plus the loss of the extra allowances for whatever appointment he would have held in India, plus the higher cost of living in the United Kingdom, which involved an officer in spending between £200 and £250 pa more than his actual pay, effectively discouraged Indian officers from applying.

Accordingly, few officers in India at the beginning of the twentieth century received any military education after leaving Sandhurst. The Sandhurst syllabus, in 1899, consisted of courses on military engineering, ballistics, elementary field works, topography, tactics, military law and administration, modern European languages, equitation drill and skill-at-arms. It was a useful groundwork for future studies, and promotion examinations ensured, in theory, that officers remained familiar with the practical aspects of their soldiering. But there was little incentive for officers to study more theoretical subjects, such as the principles of war, or military history, or even actual staff duties, which would have made them more fit for higher command or staff appointments.

Kitchener's proposal to set up a staff college in India overcame these difficulties. Indian Army officers no longer suffered financially, and the move to Quetta enabled them to move their horses, possessions, servants and households with no more difficulty than an ordinary change of station. A suggestion from the Home authorities that dangers might result from the development of two schools of thought, one at Camberley and one at Quetta, was countered by Kitchener with the assertion that the Army had no school of thought, and that he would be glad to see evidence of any thought at all on serious military subjects. In fact the syllabus at Quetta was tied to that at Camberley, and the examinations were equally difficult. Indian Army and British Army officers continued to share appointments in command or on the staff of brigades or higher formations in roughly equal numbers. This was not as unfair as might otherwise appear when it is remembered that although there were only one-third as many British as Indian troops in India, there were three times as many officers on the establishment of the British units as on the Indian.

Social origins of officers in India, under the Crown

The registers of the Royal Military College, Sandhurst, contain details of the occupations followed by the fathers of gentlemen cadets who studied there. In the period 1890–5, of the cadets passing through the RMC 1,180 gained commissions in the British Army and 129 in the Indian Staff Corps, while 58 did not complete the course. Of these, some resigned, some were asked to leave, and three were drowned (one in the College lake).

Nearly half of those who entered the Indian Staff Corps had some family connection with the country. Only about 10 per cent (113) of the cadets joining the British Army were sons of Indian military or civil officers, 72 of them from the well-paid Indian Civil Service. (In contrast only nine sons of ICS fathers entered the Indian Staff Corps in this period.)

By comparison, over one-third of the cadets who entered the British Army had fathers who were British Army officers. Their ranks included one field-marshal, 124 generals of various grades, 223 lieutenant-colonels or colonels, one cornet and one ensign (both retired). The 39 Royal Naval fathers included 9 admirals and a retired midshipman.

No member of the nobility sent his son to the Indian Staff Corps during this period, and only one knight did so. The British Army, by contrast, recruited four sons of earls, four sons of viscounts, six sons of barons, and eight sons of baronets. A further five fathers have their occupation listed as lord, but as this includes among them Lord Randolph

Churchill, father of Gentleman Cadet Winston Churchill, this term clearly includes the younger sons of dukes, who were not peers of the realm. Nevertheless, it is clear that the social status of the officers of the Indian Staff Corps was lower than that of the British Army as a whole. The officers going to India were drawn, as in the earlier period, predominantly from the less affluent families of the British middle class (see Tables 13 and 14).

Table 13

Class origin of Indian Staff Corps officers, 1890–1895

(Occupations of fathers of those cadets at RMA Sandhurst, 1890–5, who joined the Indian Staff Corps)

(a) *Fathers employed in India*	*No*	%
Army Officers (Indian Staff Corps)	29	24·1
ICS	9	6·9
Others	2	1·5
TOTAL	40	

(b) *Fathers resident in the UK*	*No*	%
Army officers	57	44·1
Naval officers	5	3·8
Surgeons	5	3·8
Private gentlemen	4	3·1
Clergy	4	3·1
Civil engineers	2	1·5
Gentlemen farmers	2	1·5
Doctors	2	1·5
Others	8	6·2
TOTAL	89	
GRAND TOTAL	129	

War emergency officers

With the great expansion of the Indian Army during World War I, officers had to be recruited from a far wider social range. Temporary commissions were granted without the long screening and selection processes involved in the Sandhurst course, and men received only the briefest training in the duties of an officer. Inevitably there were criticisms by the regulars of some of the temporary officers, aimed as much at their social as their military shortcomings. In 1917, Colonel Maunsell of the Scinde Horse was astounded to find himself, after detraining at a rest camp in northern Italy, being advised by the Assistant

Table 14

Class origin of British Army officers, 1890–1895

(Occupations of the fathers of cadets at RMA Sandhurst, 1890–5, excluding those cadets who joined the Indian Staff Corps)

			%
Army officers: British Army	540		42·8
Indian Army	37		2·9
Navy	39		3·1
Total in the services		616	49·6
Private gentlemen		170	13·7
Clergymen		86	6·9
Indian civil service		72	5·8
Others employed in India		4	0·32
Judges	7		0·56
Barristers	27		2·1
Solicitors	29		2·3
Stipendary magistrates and recorders	6		0·48
Total in the legal profession		69	5·5
Merchants		36	2·9
Medical profession		23	1·8
Civil servants		19	1·5
Engineers		17	1·3
Bankers		10	0·81
Brewers		9	0·72
Manufacturers		8	0·64
Stockbrokers		8	0·64
Landowners		6	0·45
Teachers		5	0·41
Shipowners		4	0·32
Diplomats		4	0·32
Farmers		4	0·32
Tradesmen		4	0·32
Land agents		4	0·32
Others		60	4·8
	TOTAL	1,238	

Provost Marshal not to get drunk, and not to visit the local brothels. He was even more astounded to find that half a dozen of the ninety-odd officers and gentlemen to whom this appeal was made actually appeared drunk when the time came to board the troop train, and were not even placed under arrest.

Captain G. F. Paterson, a regular officer of the 34th Sikh Pioneers, wrote home to his wife from an officers' holding camp in Sinai in September 1918: 'I have got quite a good

Bullock draught of Royal Artillery 40-pounder guns. Photograph c 1880

Royal Artillery 40-pounder guns firing. Photograph c 1880

fellow in with me, which is very lucky, because there are some extraordinary things in the way of Officers in this camp, a lot of them going to the Indian Army Reserve, no wonder the Natives think they are fit to have commissions when they see some of the specimens sent to be officers over them.' In November 1918, having been transferred to the 32nd Pioneers he wrote: 'One or two of the Indian Army Reserve fellows in this regiment are rather quaint, typical ex-tommies of rather low order with the ordinary language they use in normal conversation to match.' The regimental officers' mess of this unit, he wrote, was organised by the regimental medical officer, 'who is a Native, a Parsee'. During World War I qualified Indian physicians were commissioned as medical officers in considerable numbers, although their services were not always gratefully received by the more colour-conscious of the British officers. Paterson complains that this doctor was swindled by the mess servants and that the officers thus had to pay more for their messing than would otherwise have been the case. 'I have gone for him once or twice about it, and all he says is "What can I do". Typically native!'

Indian commissioned officers

The social class that was never recruited as officers, until after World War I, was that of the Indian gentleman. Despite Queen Victoria's proclamation, at the time of the assumption of direct control of India by the British Crown, that 'no person by reason only of his religion, place of birth, descent, colour, or any of them, shall be prevented from holding any office, the duties of which they are qualified by their education, ability and integrity to discharge', no native of India was selected to rise to commissioned rank for another sixty years. In part, this was a matter of security. The Indian Mutiny was never really forgotten by the British in India, and there was considerable reluctance to see military power of any sort in Indian hands, or any other kind of power for that matter. Another problem was that of job opportunity. Any young Indian appointed to an officer's post meant one less available to be filled by a young Englishman, and we have seen how the limited number of vacancies each year were eagerly competed for by candidates, often from families historically connected with India, who lacked the means to enter the British service.

A further difficulty was the suitability for officer posts of the possible Indian candidates themselves. The so-called martial races, from whom the bulk of the soldiers were recruited, were also the most backward in terms of western education and intellectual achievement. This was, at least in part, why they were deemed to be 'martial', because they were so much the more amenable to military discipline and the rigours of campaign. Even the dullest British officer received further education in academic and military subjects at Sandhurst, to enable him to grasp the rudiments of his profession and to lay the foundation on which his future career could be based. But the average young man of the martial classes of India had no such education, and the traditional military distrust of academic work was carried to extremes. Even basic literacy and numeracy were regarded as the hall-marks of the *babū*, the despised clerkly class.

The only class in India which did provide itself with the necessary education was that of the *babū* himself, or rather the middle-class urban intellectual.

Thus the only choice seemed to be between the martial classes who lacked learning, and the learned classes who lacked the martial qualities of leadership, physical prowess, and discipline. Yet both sides pressed the British hard to grant their sons commissions. The martial classes argued that they had loyally supported the British regime in the past, and that the British had a moral obligation to reward them for the lives and blood they had lost in the British service. It would be prudent, they said, to ensure their future loyalty by admitting them to the officer corps. The urban intelligentsia, associated with the growing Indian nationalist movement, argued that India was one nation, not a collection of communities. The non-martial classes, in their view, contained as many men with officer qualities as the martial, especially as the growing complexity of modern warfare called for skills and techniques which could be imparted by training rather than heredity.

The defeat of the Russians by the Japanese in 1905 showed Indian public opinion that Asian officers could acquire and use the skills of western military education. India's military traditions were no less strong than those of Japan, and it was clear that, given the opportunity, Indians could make just as good officers as the Japanese.

It was strongly argued by supporters of the martial-class theory that soldiers of these classes simply would not obey Indian commissioned officers of non-martial background or even of martial classes other than their own. Moreover, there were cases where sepoys had failed to respond even to British officers from another regiment temporarily set over them. Colonel Maunsell of the Scinde Horse outlines the thinking of such men as: 'This sahib is not our sahib; he has neither the gift of patronage nor the power of punishment so we need not bother about him.' In fact, during World War II, although difficulties did arise when NCOs from the old martial classes were sent to staff newly raised units recruited from traditional 'non-martial' classes, Indian commissioned officers proved just as well able to command men of different races as their British counterparts, and in some ways better able to establish a *rapport* with their fellow-countrymen.

Other objections were raised of a purely social nature. Indians married much earlier than British officers, and would thus have less time to give to their men. Their wives might speak only Hindustani or other 'native' languages, and so would not fit in with the British wives. Indian officers would demand 'native' food in the officers' mess (although the affection for his curry of the British Indian officer was a noted fact). Indian officers, with families to support, would get into debt. This was true, but ignored the fact that, despite their higher pay, most junior British officers of the Indian Army were in debt to the regimental (Indian) moneylender. Indian boys of good family, whether urban or rural, were brought up surrounded by indulgent relatives and obsequious servants which sapped their moral fibre and capacity for self-reliance. This also was true, although not much more so than was the case in some wealthy British families, and Indian parents with military ambitions for their sons found the same solution in sending their sons to boarding schools where the virtues of healthy exercise of mind and body were both preached and practised.

'Indianisation' of the Officer Corps

In 1917 ten places at the Royal Military College, Sandhurst, were set aside for suitable Indian candidates. First priority was given to the sons of servicemen, and the boys selected were mainly from military families, or the sons of wealthy landholders or Indian princes. They were the most conservative and politically reliable, ie, inert, group that could be selected, and they followed exactly the same course as British cadets. Insufficient care was taken to prepare them for life at Sandhurst, in a foreign country far from home. Of the eighty-three Indian cadets at Sandhurst between 1918 and 1926 about one-third failed to complete the course. Those who did were gazetted as King's Commissioned Officers, drawing the same pay and allowances as their British colleagues. It was ironic that this group, the most loyal to the British, was the one to incur the first shock of contact with the British community in India. As units moved from station to station, the same embarrassments repeated themselves, about such matters as admission to the station clubs, hitherto reserved for Europeans only.

At first the career opportunities for Indian and British officers were the same. But in 1923 it was decided to select eight units for 'Indianisation'. This concentrated Indian officers into a few regiments. The British officers of these corps tended to feel themselves rather degraded, and few sought to join them. The Royal Military College Magazine informed its readers: 'You are assured that you will never be asked to serve under an officer of another race.' In 1927, to cope with the increased number of Indian officers, the posts of ressaldar, subedar, and jemadar in these regiments were abolished, and Indian commissioned officers became troop or platoon commanders in the same way as subalterns of British regiments. However, it placed them at a disadvantage compared with British subalterns of Indian regiments and reduced their chances of obtaining regimental command from one in three to one in eight.

Most of the first generation of Indian commissioned officers rose to high rank in the armies of independent India and Pakistan. They have preserved the good traditions of the British officers whom they replaced and by whom they were trained. Present-day officers of the South Asian armies almost seem more British than the British, and have adopted many of the mannerisms in speech and behaviour of their British predecessors. Although the bulk of the soldiery and many officers still come from the traditional classes, men from urban backgrounds, who proved their military competence during World War II, compete on equal terms. Although the old martial classes are no longer educationally backward, the urban classes have the right to bear arms. Some of India's most successful commanders have come from communities which the British would not admit to military service, yet in their behaviour and speech and military skill they could scarcely be told apart from British Indian commanders.

The Indian Military Academy at Dehra Dun continues to give its students an education comparable with other national military academies, and the young men there are still called 'gentlemen-cadets', a title which has been discontinued at Sandhurst as a vestige of feudalism and replaced by the more egalitarian 'officer cadet'.

Social life

The social institution to which all officers of Indian or British regiments belonged was their own regimental mess. These were not mere dining-clubs, or accommodation centres, although certainly that was where officers dined and where their quarters were provided. The mess was a body corporate on its own account, having its own property, rules, and customs; acting as the custodian of regimental traditions, it was a tangible focus for the loyalties and *esprit de corps* of the officers. Unmarried officers took most of their meals in the mess, married officers only those on formal occasions, except on service. In peacetime single officers shared large bungalows in groups of two or three, and married officers had their own establishments.

Officers were discouraged from marriage before they reached field rank; 'subalterns do not marry, captains may marry, majors should marry, and colonels must marry' was the usual formula. But under the Company's rule there were few Englishwomen to marry in any case. Many daughters of Indian military officers married other military officers. It was common to find two or three sisters in a military family marrying officers. In one case six sisters all found military husbands. Widows of officers also frequently found consolation by marrying other officers in India. It is true that there were few Europeans except military men in India for them to marry, but certainly no eligible woman lacked for suitors. The climate, disease, unsuitable clothing, insanitary conditions, and the primitive state of medical knowledge at the time all contributed to a high mortality rate. Many officers were bereaved more than once, some even marrying a third or fourth time after tragic losses of earlier wives and children.

Furlough and privilege leave

Officers of the Staff Corps were allowed a furlough of one year in four. The advent of the steamship, the railway, and the opening of the Suez Canal in 1869 shortened the journey time between England and India to about three weeks, and it became far easier for officers to return to their own country. Many married while on furlough, bringing their wives back to India with them. In many cases their wives were in some way connected with Indian families, partly because the social conventions of the time tended to restrict the number of eligible young women that an officer met to those of his own circle (the girl-friends of Sandhurst cadets at this time were frequently their own cousins, or the sisters or cousins of other cadets), and partly because the life of an Indian Army officer's wife would not appeal to anyone who had not some idea of what was involved. Shipboard romances were common. So many girls caught husbands from the Indian services on the way out to India that the P. & O. steamship company became known as the fishing fleet, and the British India shipping line was called the Bibi line – *bībī* being the Hindustani word for a lady, and coinciding with the company's monogram.

Privilege leave in India was allowed of up to sixty days a year, or ninety days for officers in frontier or other distant stations. This could be used in whatever way an officer desired. Those seeking European company, with hotels, theatres, dances, and similar social functions, would go to a 'hill station' in the foothills of the Himalayas or in the hilly regions of southern India. Simla was the official hill station for the Bengal Government, Poona for Bombay, and Ootacamund for Madras. There the cooler temperatures allowed Europeans to escape from the heat of the plains and live a more active life. Central and provincial governments moved into them for five months of every year, and the whole machinery of administration, files, documents, clerks and all, was virtually given a nomadic existence. Officers sent their families to the hill stations for these five months whenever possible, but being able to join them for only eight weeks or so had to leave them as 'grass widows' (this is the origin of the term) for the rest of the time.

A great deal of flirting and philandering, and infidelity, was said to go on at these hill stations. Every season there was scandal of greater or lesser degree, and always someone lost his or her reputation or incurred disgrace of some sort. Simla in particular had a reputation for excessive gaiety. Sir O'Moore Creagh, Commander-in-Chief, India, from 1909 to 1914, noted in his *Memoirs*:

> Dinner parties, dances, and theatricals were of nightly occurrence, and in the afternoons there were race-meetings, polo tournaments, dog-shows, etc, without end. Stupid gossip or shop replaced intelligent conversation, and the subservience to superiors was truly oriental. When the big man gave his opinion nobody dared to differ. High officials full of pride sometimes mistook want of politeness for dignity, and their ladies were in a perpetual state of anxiety concerning their precedence. There was such extravagance that many of the officials were either in debt or could only just scrape along.

Colonel S. Dewe White, of the Bengal Army, writing in his *Travels in India* (1844–66) remarks:

> An officer aspiring to get a civil or military appointment, who desired to get to some place where, by currying favour with the great, he might create an influence for himself sufficient to secure that object, would select Simla . . . Here the place-hunter would stick during the whole period of his leave, taking every opportunity to ingratiate himself with all who could do him a good turn.

It was even suggested that some officers in search of staff employ, but lacking interest in high places, sent their wives to obtain it for them.

However, such episodes were not always one-sided. One subaltern, found guilty of behaving disrespectfully to his colonel by seducing the colonel's wife, was ordered to be dismissed from the service in disgrace. The Commander-in-Chief, Sir Charles Napier, looked at the papers and decided that the fruit which the officer had stolen had not required much shaking. He quashed the sentence, remarking that history recorded but one Joseph.

Officers might find other distractions on their leave. Big-game hunting was popular, involving not only the skill and excitement of tracking and stalking the quarry, but long journeys into remote forest or mountain regions and contact with the unsophisticated local people. Vast quantities of larger mammals and birds were slaughtered by military sportsmen, in some cases to the benefit of the villagers and their crops, and in some cases to the detriment of the existence of species. Officers' Mess ante-rooms and private bungalows were, in extreme cases, vast hecatombs containing the skulls, horns, and stuffed heads of victims of British marksmanship. In the climate of opinion at the time, the collection of trophies was a perfectly proper, even admirable, pursuit. Indeed, prior to modern inventions such as the telescopic camera or the game park, this was virtually the only way of physically examining specimens, and many additions to scientific knowledge were made in this way.

Other sports and pastimes could be had without long leave. Mounted sports were popular. The pursuit of wild pigs by spearmen on horseback, or 'pigsticking', was an exciting, even dangerous, and popular sport. The Peshawar Vale Foxhounds gave the garrison officers a chance to hunt jackal instead of foxes. Polo was codified and regularised, and became popular with cavalry and infantry regiments alike, British and Indian. Indeed, the efficiency of individual officers and of whole regiments was judged, it was widely believed, more by their performance at polo than by their fitness for battle. From the latter part of the nineteenth century organised team games were exported to India from British schools. Hockey in particular was popular with the sepoys, and the officers took part in the same teams as their men. There is no doubt that the opportunity to take part in field sports was a factor in making Indian service more attractive than it otherwise would have been.

In the earlier part of the nineteenth century, without such distractions, the more usual pastimes of officers were the traditional military ones of drinking, duelling, and womanising.

The arrival of increasing numbers of Englishwomen of their own class, from the 1850s on, with the increased moral rigidity of the Victorian age, brought social disapproval of liaisons between officers and Indian women.

Duelling, in India as in England, was always officially condemned, and counted as a criminal offence, the worse for being premeditated. There were few enough British officers in India as it was, without their being allowed to kill themselves. However, it was not considered dishonourable among officers themselves, and sudden deaths could always be attributed to natural causes. The climate of India seems to have bred a certain intolerance of spirit in the British officers there. Military men are conditioned to instant obedience from those below them, and those with long service in India, surrounded by servants, in a community where they were used to exercising authority, were especially prone to resent any opposition. There are several cases recorded of intemperate behaviour leading to seizures or paroxysms in argument. The caricature of a retired 'Poona colonel' exploding in wrath at some minor irritation in his club in London or Cheltenham is based on at least a modicum of truth.

Financial problems and pensions

No one who has studied any quantity of either official or private papers connected with officers of the Indian Army can fail to notice how pecuniary considerations are a constant problem, running through the whole story, like the metal strip in a bank-note. Captain Donald Stewart of the 9th Bengal Native Infantry, writing to his young wife in England in 1856 and 1857, has much the same preoccupations as Captain George Paterson of the 34th Sikh Pioneers writing to his young wife in England seventy or so years later.

> I have been appointed to the lucrative post of Station Staff for which I am to get the enormous sum of 20 rupees (£2) a month. (*April 1856*)
>
> I might get something to do at home, an adjutancy of Militia is not a bad thing. I hear it is upwards of £200 a year with quarters. If it is as much as that I might cut the Service and retire to my own fig-tree at once – I mean as soon as I am entitled to a pension. (*11 May 1856*)
>
> I shall not stick in the Service longer than I can help for the idea of dangling on in a subordinate position, for upwards of thirty years is something awful to contemplate. I have an idea that I shall be lucky if I am spared; but if I get the steps I anticipate I shall have to pay heavily for them. (*24 May 1856*)
>
> I have never been so poor as I am at present. When I come to think, I almost despair of ever being able to go home. (*August 1856*)

All my hopes of promotion have been knocked on the head ... No new corps are to be raised. (*September 1856*)

This sad blow [the mutiny of his regiment] has ruined me, and I shall now be unable to go home [on an expected furlough] even were I permitted so to do by Government. (*29 May 1857*)

I won't be a major after all, and will probably have to wait another three years for it. I don't care a blow about the rank but it is rot [*sic*] being done out of the extra pay which would just make all the difference to us. It is just like the stingy Indian Government though and it makes me more fed up than ever with the army. I'd chuck it tomorrow if I could get a better job ... The Indian Government are the stingiest crowd I have ever met and do anything to make money off the British officer. During the war, Sepoys have been drawing more than double the pay they got before as well as being given free rations and clothing, while all they have done to the British Officers is to stop our Exchange compensation so I get about two pounds a month less pay than in peace time. (*16 January 1919*)

Retirement pensions varied according to the rank and length of service of officers, as is the usual way of such arrangements. Under the Company's rules every officer who had served twenty-five years in India (including one furlough) could retire on the full pay of his rank, computed at infantry rates. Captains and field officers retiring on medical grounds, no matter what the length of their service, received pensions on half-pay of their ranks, as did lieutenants with more than thirteen, or cornets, second lieutenants and ensigns with more than nine years' Indian service. Subalterns with six years' Indian service could retire on medical grounds with the half-pay of an ensign.

These rules, although allowing officers to retire without loss of pay at the age of forty-three or thereabouts, did not oblige or even encourage them to do so. On the contrary, because of the substantial difference one promotion would make, men stayed on as long as they saw any chance of obtaining it. The average time for a cornet or ensign to reach the rank of lieutenant was seven years, and from lieutenant to captain, fifteen years. An officer would therefore be about forty on obtaining his captaincy, and might well be prepared to serve on until fifty to become a major. Many men took their pension, plus the lump sum subscribed by their brother officers, and retired on reaching that rank, but those who stayed on could expect to become lieutenant-colonels after another six years or so, and the prospect of the rewards to be gained for this comparatively short time retained many elderly officers on the active list, as there was no compulsory age for retirement.

Officers, under the Company, were entitled to furlough, or long leave out of India, after ten years' service. Until 1854 the leave period was for three years, counted from the time an officer left his unit until the time he rejoined it. He was paid for not more than two-and-a-half years out of India, at United Kingdom rates, which varied from £1 12s 8d per claim for colonels of cavalry to 5s 3d for ensigns of infantry. Extension to furlough was allowed, without pay, if recommended, in the words of the Regulations, 'by at least two gentlemen, eminent in the medical profession'. Officers who were obliged to return to Europe on medical grounds were allowed to do so, even if they had not completed ten years' service, and so were officers obliged to return to Europe on urgent personal business, although in the last case only one year's leave was allowed, without pay.

To supplement the official benefits, contributory funds were set up by the civil and military officers in each presidency. Officers paid a lump sum on joining and on each subsequent promotion, rates varying from Rs 4,350 to Rs 723·12 for a colonel on joining the scheme and on promotion respectively, and from Rs 450 to Rs 150 for a lieutenant. The monthly subscription for a married colonel was Rs 62·8, for a bachelor colonel Rs 28·2 and for married and single lieutenants Rs 10 or Rs 5·10.

Subscribers proceeding to Europe on sick leave, but whose passage money was not admissible from government funds, could claim for the passage home and back, for self and family. If one indulgence were granted in the first eight years of Indian service, another was not allowed until after sixteen years.

Widows' pensions were payable from the fund in accordance with the rank of deceased subscribers. The widow of a colonel received Rs 238·6 each year in India, or £312 18s 0d in England, and the widow of a lieutenant Rs 71·3 or £93 8s 9d. However, a lady who had been legally divorced or separated from her husband on the grounds of adultery, or who had, as the Bengal Fund's regulations delicately put it, 'quitted his protection' – or, as they went on less delicately to put it, was 'living in a state of notorious adultery, though not divorced or separated from him by law, or who subsequent to her husband's decease may be living in a notorious state of incontinence' – were not entitled to benefit. Widows could be repatriated to Europe at the Fund's expense if no government grant was made.

Widows remarrying lost the pension, except that, if widowed again, the pension was again paid. If she had married another officer, she was not able to draw two pensions, but only that of highest rank held by either of her late spouses.

Orphans were also provided for by these arrangements, subject to an officer having made reasonable equal provision in his will between his surviving dependants. Details varied between the different schemes, but included the grant of annual pensions to children according to whether they had lost one or both parents, and according to the age of the child. Boys received the pension until they were old enough to take up a respectable profession, and daughters until either marriage or death, if they remained unmarried. On marriage, a girl received a dowry or marriage portion of Rs 1,500 (£150).

Under the Crown, arrangements continued along broadly similar lines, except that officers no longer retired on full pay but on the half-pay of their rank. As promotion was by length of service, not seniority, everyone was assured of being able to reach the rank of lieutenant-colonel after about twenty-six years' service.

Officers in battle

Nevertheless, whatever their financial worries, officers of the Indian Army proved over and over again that, given the opportunity, they could be as skilled in the exercise of their profession, and as courageous in the face of the enemy, as the officers of any other army. Paterson won the Military Cross for his personal valour at the battle of Festubert in November 1914. Of the nine British officers who led his regiment into battle only two, plus the regimental doctor, survived. 'I am in command now' he wrote to his wife, '. . . you mustn't mind if this is a bit of a rotten letter as it rather shakes one up to see all one's pals suddenly polished off. I just can't realise it, it seems so funny just three of us in the Mess now.' His battalion, 500 strong when the battle started, had 52 killed, 22 missing presumed killed, and 120 wounded. Paterson survived World War I to become a colonel and a cantonment magistrate. Donald Stewart survived the Mutiny, in which he too displayed personal bravery of no mean order, to become a field-marshal and Commander-in-Chief, India.

It is invidious to select particular examples of bravery from the achievements of such officers. Indeed, the very idea of introducing awards to officers for gallantry was at first opposed on the grounds that all officers were brave, and it was wrong to select the few who had had the opportunity of showing that they possessed a quality which they were expected to possess by virtue of their rank. Nevertheless, to give an idea of what feats were performed, the following episodes, for which officers were awarded the Victoria Cross, the highest award for gallantry in the British forces, serve as typical examples.

Lieutenant Innes of the Bengal Engineers, in the Indian Mutiny, at Sultanpore in February 1858 led a skirmishing line against an enemy battery. His party captured one gun, abandoned by the retiring enemy, but were then exposed to the next gun at point-blank range. Innes rode forward, alone, and cut down a gunner just before he could fire. Then he kept the rest of the gun detachment at bay, despite musketry fire from their infantry escort, until his own men caught up with him and drove them off.

Captain Wood, of the 20th Bombay Native Infantry, was the first officer of the Indian Army to win the new Victoria Cross, in December 1856. He was the first man to reach the parapet of an Iranian fort near Reshire on the Persian Gulf. Defending troops rushed at him and he was hit seven times by their musket fire. Nevertheless he closed with the nearest enemy leader and ran him through with his sword. Followed closely by his own grenadier company, he turned out the enemy garrison and established himself in control of the fort.

The officers of the fighting arms had no monopoly of courage. The Reverend James Adams, of the Bengal Ecclesiastical Establishment, a chaplain with the Kabul Field Force in December 1879, saw that some men of the 9th Lancers had fallen with their horses into a deep, water-filled ditch during the action at Killa Kazi. He rushed into the water, under heavy fire from the Afghans, and dragged at the fallen horses to extricate the men trapped under them. While he was helping the Lancers out, the Afghans came on at a rapid pace. He had let his own horse go in order to be of more help to the men in the water, and finally made his own escape on foot, a few yards ahead of the leading Afghans.

Surgeon Captain Harry Whitchurch of the Indian Medical Service was in Chitral Fort on 3 March 1895, at the beginning of its investment by dissident tribesmen. Captain Baird of the 24th Bengal Infantry was mortally wounded about a mile and a half from the fort, and the surgeon went out to his help. The tribesmen had already broken through the British firing line in great numbers, and darkness set in, isolating Whitchurch, Baird, and the few men of the Gurkhas and Kashmir Rifles with them. Baird was placed in a doolie, and carried by the Gurkhas until three of the bearers were killed and a fourth badly wounded. The doctor then put Captain Baird on his back and carried him. The little party's numbers steadily diminished, under constant enemy fire. Whitchurch was obliged to charge the enemy and dislodge them before he could go on, and at one point rushed single-handed at the enemy surrounding them to clear a way. Finally he brought the survivors into the fort, but Baird, having been wounded twice more, was beyond further help.

Perhaps one of the most notable acts of valour was that of Lieutenant William Kenny of the 4th Battalion of the 39th Garhwal Rifles, in Waziristan in January 1920. He was in command of a company of infantry in an advanced post which was repeatedly attacked by greatly superior numbers of Mahsud tribesmen. For over four hours he held the post against all attacks, personally engaging the enemy with grenades and a rifle and bayonet. He held the post long enough for the main force to disengage and fall back, and was then faced with the most difficult of all frontier operations, the withdrawal from a position in the face of a determined enemy. The company was hard-pressed as it withdrew, impeded by its wounded, which it carried with it. It was unthinkable in this sort of warfare to let wounded men fall alive into the hands of a cruel and pitiless enemy, where they would die slowly by the most savage tortures. To enable the rest of his company and the wounded to escape, Kenny and a handful of men went back and counter-attacked the enemy. He and his men knew they were going to certain death, and were killed fighting bravely to the last.

That men should rise to such heights of courage, and inspire equal feats from their men, from whom they were separated by race, religion, and culture, is much to the honour of the Army to which they belonged. The *rapport* between the officers, often cut off from European company and comforts, and their men, was a most successful feature of the British Indian military system. The devotion with which the *jawāns* followed their *ṣāḥibs*, leaders of an alien race from an alien country, testifies to the achievement of the officers in winning and keeping the respect and trust of their men.

8

The European Soldier

In the last resort, it was the European element of the Indian Army that maintained British control of India. No major battle was fought without a strong European element in the Army being present and playing an important part in the engagement. Physically, the European tended to be larger and stronger than the Indian. He was capable of showing greater determination in the assault, and more stubbornness in defence. In part this was because he had always been supplied with more officers than the Indian soldier. But a major reason was that the sepoy fought for pay, and had no motivation beyond fairly maintaining his side of the contract and preserving his own and his group's personal honour. The European soldier was, in the end, fighting for his country's empire. It may be true that some British soldiers enlisted for drink, that many were recruited from the lowest of the low, and that they were capable of coarse or brutal behaviour. Nevertheless every man felt himself to be altogether a superior being to any inhabitant of India, and this self-confidence (although at times leading to unjustified over-confidence) was, together with a rugged patriotism, an almost irresistible military combination. But the success of these men in the field must also be attributed to the fact that they were long-service, well-drilled, all-volunteer, regular soldiers, proficient in their trade, and happy in the exercise of it.

Indeed, only a regular army could have sustained the burden of prolonged overseas duty in a hostile environment. The British garrison of India in 1946, for instance, composed mainly of homesick, war-weary conscripts, made it perfectly clear that, after six years' fighting to make the world safe for democracy, it was in no mood to embark on another prolonged campaign to preserve British power in India. Even if the government then in power had not already been pledged to quit India, it would have been hard-pressed to hold it with the British troops then available.

Recruitment

Prior to the transfer of British dominions in India from the Company to the Crown, there were two sorts of European soldier in India, those directly employed by the Company and those belonging to the British Army temporarily stationed in India. The latter were no different from the men in other British regiments outside India, and were liable for service anywhere that their unit might be sent. The Company's Europeans enlisted for service in the East Indies only. They, like the Indian soldiers and their officers, were not servants of the British Crown, and had no great wish to be.

In the early days of the Company's army, only European soldiers were employed. Not until the mid-eighteenth century were Indian sepoys enlisted and organised along western lines. When the British Government took a closer interest in Indian affairs, it also took the power to order out to India, at the Company's expense, British troops to defend

British interests. The Company preferred wherever possible to use its own employees, but attempts to recruit men for its European regiments were constantly hampered by parliamentary feeling against standing armies in general, and those not under its immediate control in particular. It was forced to recruit by obtaining the issue of royal warrants on each specific occasion, which authorised a named individual to enlist a stated number of persons. Legislation against the admission of Roman Catholics into military service was a further inconvenience, restricting, as it did, the number of recruits from Ireland, although in fact many Irishmen did join both the British and the Company's armies.

Not until 1799 was the Company granted power to recruit and train recruits in England and subject them to military law there. Permanent recruiting offices were set up, manned by recruiting officers and sergeants of the Company's service, in London, Edinburgh, Dublin, and Liverpool. Extra offices were opened at Cork in 1822, and in 1846 at Bristol and Newry. The Company's United Kingdom Depot was first established at Newport, on the Isle of Wight, in 1801. It moved to Brompton Barracks, Chatham, in 1815, and then to Warley Barracks, Brentwood, Essex in 1843.

It remained difficult to meet the requirement for men. Many candidates were rejected on medical grounds. Others deserted after enlisting, or thought better of their decision and bought themselves out before embarkation. The British Army, itself an all-volunteer army, was always short of recruits, and competed in the same market. Because service in the East amounted to self-imposed exile, and there was a very good chance that men serving in the ranks would die from disease, accident, battle, or other violent cause, the Company's army was frequently chosen by men who wished to cut themselves off from their own world. It was common for Englishmen to enlist in the Army to evade some liability or other. Getting a girl pregnant, evading creditors, or the local constable, were just as powerful reasons for enlisting as more noble or lofty motives. For those for whom even the anonymous mass of the British Army was not sufficient to provide concealment, the Company's army was a safer refuge. As a haven for soldiers of fortune, adventurers, rogues, and petty criminals, where few questions were asked about a recruit's background, and where the authorities were not too fussy about the answers, the Company's European Regiments were very much, in their day, what the French Foreign Legion was to a later generation. They produced, too, much the same sort of hard-bitten, tough soldier. Indeed, the nickname of the 1st Bombay Fusiliers was 'the Old Toughs', as that of the Bengal Fusiliers was 'the Dirty Shirts' and that of the 1st Madras Fusiliers, 'the Lambs' – a sinister title, which other regiments in other armies earned from a similar record of ferocity.

Not only the soldiers selected Indian service to escape embarrassment at home. Captain James Crockatt, for instance, was described by William Hickey, the Bengal diarist, as 'a dissipated London dasher. He had run through an independent fortune, being finally obliged like many other spendthrifts, to seek refuge from his creditors' attacks by accepting a commission in the East India Company's service'.

A story of one of the Company's private soldiers is that of Thomas Babbage. A shop assistant in Bond Street, in 1855, he found himself impossibly in debt following an unsuccessful attempt to make his fortune by the well-known method of forecasting the winners of horse races. To extricate himself from his difficulties (he was then about nineteen years of age) he offered his services to the East India Company's recruiting office in London and the same evening was dispatched to Warley Barracks, Brentwood, where he was attested as a recruit for the recently formed 3rd Bombay European Regiment. Three days later he was on a ship bound for India. Six months later, after the usual lengthy voyage round the Cape of Good Hope, he disembarked at Bombay, having meanwhile been taken on as a surgeon's orderly (or medical assistant, as the post would now be called) in which employment he was to spend the rest of his service.

After landing, Babbage and his comrades marched to the railway terminus, where they were locked in, to prevent them from escaping and getting into mischief in Bombay city. Subsequently they proceeded, partly by train, partly on foot, to Poona where they were given their first meal (beefsteaks, onions, bread and coffee) since disembarking twenty-four hours before. On the march their water bottles were soon exhausted, and the only refreshment was half a pint of beer per man.

From Poona they marched to Mhow in Central India. At Ahmadnagar Babbage suffered 'sunstroke' through sleeping with his bare head too close to the tent walls. Recovering, he found his medical services were in great demand, when the draft was attacked by cholera. Several men died. Babbage felt an attack coming on, but treated himself by drinking a full bottle of brandy, running about as hard as he could while wrapped in as many blankets as he could carry, and then lying down sweating, with even more blankets heaped on top. He awoke feeling weak (as well he might!) but free from any symptoms.

During the Indian Mutiny he took part in the siege of Jhansi in 1858. After the assault he left the surgeon's post to go off in search of loot, collecting about £1,000-worth in an hour and a half. He was wounded in the thigh by a mutineer's bayonet, and had to return to his own medical officer to have it dressed. He received little sympathy, and was obliged to resume his proper duties with the wounded.

From Babbage's story we can pick out a number of points typical of the average European soldier's life in India. One, as we have seen, was the fact that he enlisted to avoid pressing problems of a domestic nature. Another was that of his diet, for as the European soldier was an expensive commodity, he could generally, though not invariably, expect to be well fed, and beef (to the repugnance of Hindus) formed a major item on the menu. The hope of loot was always present, although the literally golden opportunities of the Mutiny never again repeated themselves on so vast a scale. Finally, there was the incidence of sunstroke and cholera, and the superstitions surrounding their causes and treatment that medical science was still unable to combat.

Health

There is no such malady in modern medical textbooks as sunstroke. There is heat-exhaustion, caused by the body losing so much salt and moisture through sweating that it is unable to function properly, and there is heat stroke, caused by overheating of the body when the outside temperature and humidity is so high that perspiration through the skin, as a means of keeping the body at the correct temperature, does not take place. Sunburn of the skin is also a risk, and can result in a soldier being unable to carry out his duties through soreness, and blistering, through which infection can occur.

Medical science was at first baffled by the effect of the tropical sun. Men in western uniforms went down before it like flies, but it was not until the early 1840s that any attempt was made to introduce tropical dress. Sir Charles Napier, and Lord Gough, his predecessor as Commander-in-Chief, India, both wore domed sun helmets of their own design on wickerwork frames. During the wars in Afghanistan, Sind, and the Punjab, European troops had worn low peaked forage caps, often with white cap covers continued behind into a protective neck flap. However, in the 1860s the tall, domed, pith helmet, or 'solar topi', was adopted and made compulsory wear, despite complaints that they 'gave an appearance to the wearers of having been extinguished'. One reporter commented that during the worst hot weather during the Mutiny troops had marched 'actually in their shirt sleeves like a lot of insurrectionary haymakers'.

The helmet was intended to keep the head cool by interposing a space between the top of the head and the direct sunlight. It had a peak at the front to shade the eyes, and another, larger, one at the back, to shade the neck. The impact of the sun's rays on the back of the neck was thought to be a particular cause of sunstroke. It was noticed that soldiers of the United States Army operated in sunny climates in the southern states wearing small, peaked caps, without ill effect, and some authorities deduced that there were peculiarities of the atmosphere of the western hemisphere that filtered out the harmful rays. Spine pads were introduced to protect the lumbar regions, and flannel cholera belts were issued to be worn round the abdomen.

Cholera is a water-borne disease. It has been endemic in India since early times, although it did not spread to Europe until the 1830s. The disease is most commonly incurred by the victim's drinking infected water or eating raw vegetables or fruit washed in infected water. The water is infected by the body waste products of other victims, or by flies transmitting infection from these. The symptoms of the illness are especially horrible and frightening, both to the patient and to his comrades exposed to risk of the same disease. Vomiting, violent intestinal pains and continued dysentery are followed by terrible thirst and dehydration, leading to a high incidence of mortality. The low standard of hygiene in most armies, the lack of sanitation, the swarms of flies, all combined to make the risk of cholera a real and horrible one. The practice of regiments going into 'cholera camps', marching out from their garrisons into the country, on and on until the disease died out, worked in the end because, without knowing it, regiments were marching away from the infected water supplies carrying the original infection. Infection from sufferers ended as they died, and the flies, food or water from each camping place were left behind when the regiments moved on. Some idea of the effects of a cholera outbreak may be judged from that at Karachi in 1845. Before the epidemic the 86th Foot was 1,091 strong; 410 men were attacked of whom 238 died. The Bombay Fusiliers were 790 strong, of whom 221 were attacked and 83 died. Following a few sporadic cases, admissions to hospitals increased rapidly on the night of 14 June. On 15 June, 175 men, from all the eight regiments in the garrison, were admitted, and 75 died. On the 16th, 277 were admitted and 186 died. On the 17th there were 245 admissions and 116 deaths, and the next day, as the epidemic passed its peak, only 117 were admitted of whom 65 died.

Other outbreaks, albeit on a smaller scale, steadily thinned the ranks of regiments in India. The 9th Lancers, in three months during 1843, lost one officer, five sergeants, two trumpeters and eighty-three rank and file, besides fourteen wives and eight children. The 1st Battalion of the 19th Foot (The Green Howards) lost to cholera, between 1862 and 1863, one colour-sergeant, one drummer, eight-five rank and file, with five wives and eleven children; the 2nd Battalion, ten years later, lost in one year four sergeants, one corporal, twenty privates, five wives and seven children.

The disease was no respecter of persons, of rank, age, or sex. The Commander-in-

Chief, India, General Anson, died of it when marching on Delhi in May 1857, and his successor in command of the Delhi force, Major-General Sir Henry Barnard, died of it within eight weeks of him. Sir Charles Napier, the conqueror of Sind, saw cholera take his favourite nephew John, and that officer's little daughter, within two days of each other.

The disastrous effect on morale of such occurrences can easily be imagined. The sheer number of fatalities was bad enough. There was a legend among the soldiers, seeing the pattern of empty beds after an outbreak, that it was the Wandering Jew who brought the cholera with him, as he walked in a figure-of-eight. A worse horror was the terrible swiftness of the disease's onset, and the cruel sufferings of its victims. Officers and men alike saw their comrades, wives, children, literally shrivel up before their eyes, turning from living people into dehydrated bags of skin, and bone, and pain. Some victims who might have recovered if they had tried to fight against the disease, were so terrified that they made no attempt at resistance and died through fear. Soldiers who joined with romantic notions of a glorious death in battle soon learned the tragic lesson that in the Indian service there was so little fighting, so much fever.

A heroine of one such epidemic was Mrs Webber Harris, wife of the colonel commanding the 104th Foot (Bengal Fusiliers). In September 1869 this regiment suffered an outbreak of cholera, with the usual high number of fatalities. Twenty-seven men of the regiment died in a single night. The colonel's wife organised a nursing service for the regimental hospital, and arranged concerts and similar entertainments to try to distract the men's minds from the fear of infection. She even helped as a nurse in the hospital wards herself, although at this time ladies were thought to be too delicate to be subjected to such harrowing scenes. Afterwards she was presented by Major-General Sam Browne with a gold replica of the Victoria Cross, to which all the officers of the regiment had subscribed, 'for her indomitable pluck during the cholera epidemic 1869'.

The widespread nature of cholera, and many other diseases of the gastro-enteric tracts, was due in the main to the bad sanitation and poor hygiene around Indian stations. Hygiene, it must be admitted, is often, actually, a matter of opinion and custom. The Englishman thinks it a dirty habit of the Indian to blow his nose on a leaf and leave it on the tree. The Indian thinks it a dirty habit of the Englishman to blow his nose on a handker chief and carry it round in his pocket all day. Muslims and high-class Hindus are taught by their religion to carry out ritual daily ablutions that would ensure a higher standard of personal cleanliness than that of the average European. On the other hand it cannot be denied that, for a variety of reasons, the personal sanitary habits of many Indians of the poorer sort are, at least by the standards of modern north-west Europe and North America, backward in the extreme.

Armies in particular are always liable to disease from the insanitary habits of careless or indisciplined soldiers, and the flies attracted by large numbers of transport animals. The evil effects of bad camp hygiene have been known since Biblical times (Deuteronomy, Chapter 23, contains excellent advice on the subject). Nevertheless, although far-sighted commanders did all in their power to reduce the hazards to their men's health from this source, it remained, and indeed still remains, a threat to be constantly feared.

Another major threat was malaria. It was not realised until the early twentieth century that the disease was carried by a parasite of the anopheles mosquito. Quinine was used as a guard against malaria throughout Victorian times, but the disease, causing severe fevers, and often death, remained a major hazard. The connection between stagnant water and the disease was appreciated, and attempts were made to combat it by filling up all hollows around barracks, or, better still, moving troops to cantonments in hilly, better drained locations. Even as late as the 1930s, however, the number of men admitted to hospital for malaria each year was about 10 per cent of the whole army, compared to 2·5 per cent for dysentery and 2 per cent for heat stroke or heat exhaustion.

It is not surprising that the generals most popular with the men were those who did most for their health and welfare. One such was Sir Charles Napier. He had a special interest in providing the soldiers with decent barracks. He argued that the number of men in a room should be governed not by the floor area, but by the cubic capacity so that each man should have not less than 1,000 cubic feet of space. Rooms were to have painted on the door the number of men authorised to be in them, and their commanding officers were to be held responsible that these figures were not exceeded. Rooms that were a mere eight to ten feet high received his strongest condemnation: 'The stench of a low bad barrack room in the morning is appalling', he wrote. He ordered them instead to be built twenty-five feet high, with double ceilings and thick walls to keep out the heat, and good ventilation. 'Such barracks are expensive no doubt,' he wrote in his typically astringent style, 'so are sick soldiers; so are dead soldiers. But the difference of these expenses is, that the first is once and done with, the second goes on increasing like compound interest and quickly outstrips the capital.'

Napier endeared himself to his troops in other ways. One was by inviting the commissaries to sample the abominable food which was supplied to the men in hospitals, and then asking them how they expected sick men to thrive on it. Another was by attacking the luxurious ways and fine airs affected by their young officers. 'There are boys in this camp,' he wrote on one campaign, 'who require and have more luxuries than myself who am sixty-three and Governor of Scinde.' He opposed the principle of permitting regimental officers of infantry to ride on the march. Frederick the Great, Napoleon, and many other great captains had forbidden it, he said, and the Duke of Wellington had regretted allowing it, saying it cost him 20,000 cavalry in their forage alone. If one man were allowed to ride when his duties did not require it, others, who were not, would justly resent it. Soon sergeants would be allowed horses, then corporals, and finally privates, so the whole army, as in the Dark Ages, would become a mounted horde. As for other extravagances, 'the damning sin of the magnificent armies in India (Queen's and Company's) is an outrageous and vulgar luxury – I say vulgar because (like the frog in the fable) we soldiers burst ourselves in trying to live like men of £20,000 a year in landed property! We, who in private life could hardly buy a pint of beer, must drink the most costly wines!' His attempts to cut down on such behaviour made him unpopular with some officers.

Another commander-in-chief, even more loved by those who served under him, was Lord Roberts. One of his first impressions on reaching India as a young officer in 1853 was of the bad conditions in which his artillerymen lived at Fort William, Calcutta. They were crowded into small, badly ventilated buildings. Sanitation was bad and the water supply was worse. The only efficient scavengers were the adjutant-birds, so called because their rather meticulous method of walking was said to resemble that of these military staff officers. So useful indeed were these creatures that the young artillery cadets were forbidden to harm them on pain of incurring the severest censure. Peshawar, on the Punjab frontier, was another of Roberts' early postings. It inspired, he said, 'a sort of terror for the English soldier from its proverbial unhealthiness. The water supply for the first five-and-twenty years of our occupation [1845–70] was extremely bad, and sanitary arrangements, particularly as regards Natives, were apparently considered unnecessary.'

In the first half of the nineteenth century, the annual mortality rate of British troops in India averaged nearly 7 per cent. In some stations it was 10 per cent or more.

A royal commission, investigating army health in India, reported in 1864 that the annual mortality rate could be cut to 2 per cent if proper precautions were taken. Between 1866 and 1877, no less than eleven crores of rupees (about £11 million) was spent on improved barrack accommodation, and the rate by 1882 had dropped to 1·7 per cent. Much was achieved by the influence of Florence Nightingale, who was just as effective in securing reforms in the health of the Army in India as she was in the United Kingdom.

The Government of India admitted that as the British soldier was both an expensive item to obtain and a valuable one to use, it was only good economics to take care of him when he was there. Between 1834 and 1857 the Company had paid to the Crown nearly £200,000 annually for the troops stationed in India. In 1861 a capitation rate of £10 per head was introduced, reduced in 1891 to £7 10s. This was about 10 per cent of the cost to the British taxpayer of enlisting and training recruits, the pay of young officers before going to India, the upkeep of schools and training establishments, and the expenses of men sent home as invalids or time-expired. Pay and similar emoluments were paid in India from Indian revenues, and Indian revenues contributed towards part of the cost of pensions.

Drink

Both Napier and Roberts did much to oppose that other scourge of the British soldier, excessive indulgence in alcohol. This is a traditional weakness of military men, and of the British soldier in particular. Wellington held the opinion that his men enlisted for drink, and in the brutal conditions in which private soldiers lived then it is no wonder that many sought escape from their surroundings in a drunken stupor. Even more in India, the soldiers, like the officers, sought oblivion in a similar way from boredom, monotony, heat, and the ever-present fear of death and disease. Napier, however, urged sobriety on his men. Most of those who collapsed from heat stroke in his desert campaigns were intemperate drinkers, he asserted. He pointed to the advantages of sobriety, and concluded one letter to a private soldier, who had written to him, in these terms: 'Mind what you are about, and believe me, your well-wisher Charles Napier, Major General and Governor of Scinde, because I have always been a remarkably sober man.'

Roberts achieved a great deal in the cause of sobriety. He founded the Army Temperance Association, a federation of the various religious and other bodies who had hitherto preached this cause in competition with each other. There was a difference of view between those preaching temperance and those preaching total abstinence. However, Roberts was able to set up in India soldiers' institutes, where men could obtain tea and coffee, or light alcoholic refreshment, in decent surroundings, with facilities for games, books, newspapers, and writing materials. Though mocked at by the ribald, these institutes did much to improve the quality of life of the soldier, both in India and in the United Kingdom, where a similar system was instituted. Although the 'wet canteen' continued, where hard spirits as well as beer were readily available, and where those with a mind to could drink themselves insensible, from the 1880s onwards it was no longer the only place of recreation open to the soldier.

Subedar Sita Ram was much impressed by the British soldier's great addiction to rum. Some sepoys thought that it contained a sort of elixir of life. Sita Ram was convinced that the men fought simply to obtain more rum, and came to the conclusion that they would do anything for it, and that it accounted for much of their ferocity in battle. He was probably not far wrong.

Drink certainly led to the discomfiture of some of the Company's European troops at the time of the ending of their local service in 1860. There was very great discontent at the manner in which this was conducted. The men felt that as they had enlisted to serve the Company, they should be discharged, and re-enlisted for the Queen's service with the bounty paid by the Queen's government to all recruits to her army. They also objected to being transferred without their own consent, like so many chattels or pieces of property. They were, they said, Englishmen, not bullocks, to be handed from one owner to the next. Having been servants of the Company one day, if they were handed to the British on the next, would they be transferred to the Americans, or someone else, in the future?

At a Queen's birthday parade, the Madras Fusiliers conspicuously refused to cheer. Other units showed similar disaffection. The Commander-in-Chief, India, and the Military Member of the Governor-General's Council both sympathised with the men, and thought they had a genuine grievance. However, the Governor-General, supported by the Secretary of State, remained obdurate. Why should the European soldiers be paid a bounty when the native soldiers were not? The answer was that the Europeans were now being asked to accept service out of India. However, the logic of the Treasury prevailed, and no money was forthcoming.

Meanwhile the discontent increased. Although referred to by wits as 'the Dumpy Mutiny' (from the fact that during the crisis of the Indian Mutiny the height standard for recruits had been relaxed) the threat to the British position in India was quite a real one. The Mutiny had only just been quelled, and the country was still in a state of excitement. There was a large-scale campaign in progress on the North-West Frontier. As a compromise it was decided that men who did not wish to serve under the new conditions might take their discharge, and some 10,000 disgruntled men, about two-thirds of the whole force, actually did so. The loss of artillerymen was particularly great.

In order to keep as many men as possible, the colonel of the 3rd Bombay Europeans hit upon the plan of making the canteen free to everyone. After two nights men were in such a state that they were declaring their willingness to serve the Queen under any conditions, or even no conditions at all. They were accordingly induced to sign the forms of transfer. The bar was shut, and next morning revellers were confronted with the undeniable evidence of their own signatures or other marks. This was rather discreditable to their officers, as less than half led their men into the British service, the majority preferring the more lucrative terms offered by the new Indian Staff Corps. Our old friend Thomas Babbage was in this regiment, but being a prudent man he had partaken sparingly of the proferred refreshment, and so resisted the temptation to sign. He and the others were sent home in the most ill-found, ill-victualled ships the Government could find, it being firmly believed by the men that this was intended as revenge for their behaviour. Many of them promptly re-enlisted into the British Army, but Babbage remained in civilian life.

EARLY NINETEENTH CENTURY

One of Babbage's fellow soldiers before the walls of Delhi was Private Frederick Potiphar of the 9th Lancers. He was an experienced soldier with twelve years' service to his credit, and subsequently wrote a narrative of his experiences in the Mutiny campaign. Most of this is concerned with the tactical operations, or with Potiphar's activities in fighting with or wreaking vengeance on the mutineers. Nevertheless there are some interesting details of the daily routine of a cavalry regiment on the line of march.

Moving to join the army before Delhi, the regiment halted each night at caravanserais, resting-places built for the convenience of travellers by pious Muslim rulers of former days, each spaced at intervals of one day's journey. 'The first thing wee done,' he writes, 'as is the custom of all Dragoons, was to see to the feeding and watering [of] our horses, this done wee then began to look to ourselves . . . Our followers soon began to come up and were soon at work to get us something to eat. The rations being issued wee set to and soon got a meal ready, by this wee felt ourselves refreshed and also a tot of liquor.'

An evening issue of rum or other spirits seems to have been normal practice in this regiment, at least on active service, and was no doubt a great comfort to men who had spent all day riding or leading their horses.

The usual routine in any hour was ten minutes walk, ten minutes trot, ten minutes walk, ten minutes trot, then dismount and lead for ten minutes, and finally ten minutes halt.

Cholera and dysentery took their toll of Potiphar's regiment as of others. He mentions the trumpet-major as one of the first casualties. The doolies provided for the conveyance of the sick he describes as 'not altogether a pleasant affair to ride in'. On the other hand, visiting some of the sick and wounded of his regiment after they had marched to the relief of Agra, he notes 'I was greatly impressed to see the attention our poor fellows were getting at the hands chiefly of the females. Some had one and some two ladies attending on them besides the medical men and other subordinates.' No doubt the natural compassion of these nurses was attended by a feeling of gratitude to the Lancers who had ridden to their rescue.

A more detailed account of the everyday life of a British cavalryman in India has been left by Private Hall of the 14th Light Dragoons, who enlisted in December 1824 and served for twenty-two years. He confesses that he was attracted to the Army by the glitter of the uniform and the promise of adventure. The recruiting sergeant to whom he applied for further information about an army career promised him a jovial carefree life, money in his pocket, and a horse to ride like a gentleman. Only after joining did he discover that the wealth was reduced by a long list of stoppages to pay for the handsome uniform, that the brutalities and petty tyrannies of NCOs prevented any gentlemanly life, and that the horse soldier's work was often dirty, sweaty, and hard.

It was not until 1841 that he embarked for India. His memories of the troopship were, by and large, unpleasant ones. The women of the regiment had to draw lots for the limited number of vacancies on board. There was no 'points system' and all had to enter the lottery, irrespective of the length (or legality) of their marriage, or of their husband's service in the regiment. There were 400 soldiers on board, plus the fortunate women and children. The other families were almost certainly parting for at least ten or fifteen years, probably for ever. The soldiers were allowed twenty-two inches' width of deck each for sleeping quarters. Officers had a cabin each, and the sergeants and their wives a separate section in the forecastle. Grog and porter were issued, and fresh meat for a few days, but then salt pork, salt beef, and peas. By custom the cooks were allowed to skim the fat from the large coppers in which the meat was boiled, and keep it in some of the empty casks from which the meat came. 'There is a great deal of fat to be saved during a long voyage like this . . . and at the end of the voyage this fat sold pretty well. I am sure no one need envy them their situation, for I would not have done it the whole of the voyage, for no consideration, for it was enough to kill almost any man, in some of the climates we had to pass.' It must be remembered that Hall and his comrades were travelling in a sail-driven East Indiaman, going round the Cape of Good Hope. The sailors, he noted, during the routine task of pumping the ship's bilges, sang 'some of the most vulgar songs during that time that I ever heard in the whole of my existence, shameful to be uttered from the mouth of man'. Such songs, to have shocked even a cavalry trooper like Hall, must have been ribald in the extreme.

The 14th Light Dragoons arrived at Bombay in early September. After disembarkation, and an unlooked-for incident when one of their troopers was recognised as a deserter from the Bengal Horse Artillery and sent back to join his old corps, they received an issue of troop horses and rode to Kirki, seventy miles away, where the 4th Light Dragoons, whom they were to relieve, had prepared a friendly welcome for them. They were met and played in by the regimental band. The barrack rooms were laid out ready to be taken over. Food was set out on the canteen tables, with a large choice of dishes. Such arrangements, indeed, were no more than a well-administered unit could be expected to organise, but an unexpected touch was the supply of two drams of arak (local spirits) for each man, paid for by the troopers of the 4th out of their own pockets.

The men of the 4th were obviously in good spirits at the prospect of going home, and gave away many items of the white cotton hot-weather uniform, for which they had no further use, to the troopers of the 14th, who would otherwise have had their pay stopped to buy a new issue. Each man needed eight or ten pairs of white trousers, half a dozen white jackets, a dozen shirts, and several pairs of white socks. Bedding was another item for which the troops' pay was docked. It consisted of a hair mattress, two rugs or blankets, and two sheets, together with a wooden cot. Hall found, at first, that even a sheet was more than enough 'as it is so very warm, perticuarily for the European soldier on his first arrival in this country, and the musketors and bugs that are so numerous out here . . . will not let one have but very little rest for some length of time after a man comes out here, so long as there is any European blood left in a man as the saying was'.

Several seasoned troopers of the 4th Light Dragoons volunteered to transfer to the 14th and stay on in India. Hall remarks that they all subsequently said they regretted their decision, which was taken partly from the attraction of a substantial bounty (which was, however, soon spent) and partly under the influence of alcohol, as the regimental canteen was left open, and drinks were served free, on the day set aside for men to volunteer for continued Indian service.

Hall gives a very detailed description of cantonments in which his regiment was quartered. They were quite different from barracks in the United Kingdom in that there was no perimeter wall to confine the soldiers. Apart from a few forts, everywhere was as open and airy as possible, and there was plenty of water, with enough troughs to water the horses of a complete troop all at once. There was a level plain, stretching for two or three miles, where the regiment could go through all its exercises and gallop freely (a refreshing change from the restricted countryside at home). There was a native bazaar where the men (paid Rs 15 per month) could buy whatever personal things they needed. On one side were the huts for the Indian grooms and grass-cutters allotted to the regiment, and near these the farrier's shoeing shop, the veterinary officer's surgery, and the horse hospital, or lines for sick horses.

The regiment's horses were kept in long lines in the front of the station, sixteen lines, two per troop, with each horse standing three yards apart from its neighbour, protected by a horse blanket in bad weather. The common practice, although Hall does not mention it, was for the two lines of horses of each troop to face each other, separated by a broad avenue, where men and equipment could be lined up for inspection. The 'lines' were long rope cables lying along the ground, to which the horses were tied by head ropes. At the rear of the horses were lines of picketing pegs, to which the horses were tied by heel-ropes. Similar arrangements were made for the horses of artillery batteries.

Behind these was the Main Guard, or guardroom, a large building, standing apart, containing the armoury and treasure chest; then eight bungalows, all in a line, one for each of the troop sergeant-majors. In echelon behind these were sixteen large barrack rooms each holding half a troop, with spacious latrines on either flank, some distance away; then all the 'usual offices' of a regiment – the Roman Catholic chapel, the cookhouses, the armourer's shop, the canteen, the regimental sergeant-major's bungalow, the schoolroom, the quartermaster's store, a theatre, a ballroom, tailors' and shoemakers' shops, the picquet hut, and the rear guard house. On either side of this complex were the married men's bungalows. Well to the rear were the officers' compounds, usually occupied by only one officer, married or single. These compounds covered two or three acres, and many had flower gardens and hedges of prickly pear around them. In the middle of these was the officers' mess, and the garrison church.

The regiment had its share of cholera cases, and of murders and suicides committed by soldiers driven mad by the heat and boredom of garrison life, a syndrome that was to remain with the British Army throughout its occupation of India, as with the armies of other European nations in their colonial stations.

Hall was on active service during the First Sikh War, and although surviving several encounters with the enemy, was eventually laid low by a severe pain over his heart. He was bled in both arms, which relieved the pain for a time, but it soon returned. Two dozen leeches were then applied to the seat of the pain, and the efforts of these creatures seem to have had some success, as Hall felt a little better. But the regiment, still on active service, had to move on, and its sick and wounded had to move with it.

> I was ordered to be carried in one of the hospital palanqueens as mine was a very severe case. It is very trying in sickness on the line of march in a Foreign land, as there is very few comforts for the private soldier, only just the hospital allowance. If you cannot take that, there is nothing more to be got.

Despite the discomforts of travel by palanquin, or doolie, Hall counted himself more fortunate than some of his comrades who had to ride on the 'hackeries' (ox carts). He was able to mount his horse again after a few days, but never again felt really fit and in 1847 was sent back to England and invalided out after twenty-two years' service. He had throughout his Indian duty taken advantage of a scheme enabling him to remit part of his pay to an account kept for him by the pay authorities in England. He confesses that he had never really expected to return to spend it, but thought it might make some useful contribution to a relative. Now he was able to collect a useful lump sum, and lived to a good age in retirement, a model of temperance and piety.

LATE NINETEENTH CENTURY

An account of life in the ranks of India later in the century survives in the memoirs of John Fraser of the Northumberland Fusiliers. These were published in 1939 as *Sixty Years in Uniform*. Fraser joined up in 1876, to escape from the unemployment endemic in north-east England. Coming as he did from a respectable working-class home, his decision cast his parents into gloom, for to enlist voluntarily in the Army was always a matter for shame. Popular opinion was convinced that only bad lots joined the Army, men who had some reason for wishing to disappear from view. His posting to India caused them even more distress, being regarded almost as a virtual death sentence.

However, he found the barracks at Agra, where his regiment was posted much more spacious than those in the United Kingdom. He describes them as having an affinity in size and shape to a cathedral. The place of the transept was filled by the mess room, fitted with tables and forms so that the whole company occupying the barrack room could sit down to eat together. The two halves of the main room thus bisected were used as sleeping accommodation, being about twenty feet wide and fifty feet high. The walls were interspersed with arches, and the floor was made up of stone slabs. Each arch contained a shelf, a narrow iron bed, and a square metal kit box, corresponding to the modern locker. Outside these arches ran corridors, about fifteen feet wide, with doors opposite each bed, and outside these doors a veranda, the roof supported on the outer edge by a row of pillars. From the veranda, a plinth of five steps on all four sides went down to road level. The buildings, built about 1860, were of light-coloured stone, limewashed once a year, and the whole camp was designed to provide barrack rooms, stores, and offices for a complete infantry battalion.

The barrack rooms were cooled by a 'punkah-wallah' (*paṇkhā vālā*) who by pulling on a rope activated a large swinging cloth fan in the roof. He worked through the hottest part of the day and night, stopping only in the morning between 5 and 10 am and in the evening between 4 and 9 pm. The doors on the windward side of the rooms were fitted with wicker frames, kept cool by a servant continually throwing water over them. But even in the hospital temperatures could still reach 117°F.

Nevertheless, there were compensations. All fatigues, cookhouse, sweeping, and water duties were done by native servants. For a few annas a soldier could be shaved in the barrack room by a native barber. Indeed so skilled were these barbers that a man could be shaved in bed while still asleep, or, at least, not quite awake. The rations were good, and included a pound of meat per day, plus a pint of beer and a dram of rum. Fraser records sadly that heavy drinking was then common both in the Army and in civilian life.

Fraser himself was a temperate man, and a man of some intelligence. He soon rose to non-commissioned rank and before long was a colour-sergeant, in charge of the discipline, pay, accounts, and equipment of his company. He later married, and finished his sixty years in uniform as a Yeoman Warder at the Tower of London.

The Sergeants' Mess, in India as elsewhere, was a less exalted version of the Officers' Mess. The sergeants had their own dining-room and bar, billiard rooms and similar facilities. Like the Officers' Mess, it was both a club for its members and a tangible centre for their traditions and customs. It gave the ambitious private soldier a place to aim for, and membership of the Sergeants' Mess carried esteem as well as valuable social privileges. When regiments were in India, parties were given monthly to which local European residents, their wives and daughters were invited. Many Europeans in subordinate civil capacities, on the railways or in the telegraph department, were former NCOs of British regiments who had taken their discharge and settled in India.

The hot weather of the Indian plains proved a trial to Fraser and his comrades. Attacked by prickly heat, they could do little more than lie and wait for the dramatic arrival of the monsoon. At last the great rain clouds rolled in, the temperature fell, and men felt the great relief of being able to breathe again. Not for long, however. The rain brought floods, and the floods brought a chorus of frogs croaking through the night. Mosquitoes, vicious and numerous, emerged to torment soldiers in their barrack rooms. Flying ants, attracted by the oil lamps, rushed to their room in such numbers that their charred corpses extinguished the flames. They were joined by large flying bugs, which burnt with a particularly evil smell. Fever broke out, and morale dropped once more until the monsoon died away.

To occupy the soldiers in peacetime, and keep them out of mischief, their time was divided between work and play. Work consisted of drill, either mounted, gun, or foot, according to arm of the service, weapon training, and, for the more promising individuals, signalling, scouting, or other specialised training. In the technical arms, 'tradesmen' – collar-makers, wheelwrights, carpenters, etc – went about the learning and practising of their special arts. For the infantry, route marching fulfilled both a valuable training function and a political one. The target was for an infantry battalion, with its men carrying full battle equipment, to be able to march fourteen miles in a day. The regular movement to and fro, on the roads of India, of a thousand well-disciplined, well-armed Europeans, band playing, their colours in the centre of the column, field officers riding at the head, baggage train following up the rear, was undoubtedly an impressive exhibition of military power.

A well-organised regiment would have already reconnoitred its camping ground. The Quartermaster and his lascars would go on ahead of the column, and the tents would be pitched ready for the troops on their arrival. Fraser's regiment was one which ran its own mobile coffee-shop for the men. This, a forerunner of the modern 'Naafi-wagon', went ahead of the column and set itself up on the approved halting-places. As the coffee-shop came within sight, the band, with that degree of forced appropriateness beloved by military bandmasters, would strike up 'Polly put the Kettle on' or a similar air, and the men would step out knowing that soon they would have a short break and a drink. Drinking from water-bottles on the line of march was strongly discouraged. It did little to slake a man's thirst, the water being lost through perspiration, and only made him want more.

The British soldiers, like the officers of the Indian Army, had a far higher standard of living in India than they could aspire to at home. For both groups this included the opportunity for sport, and in particular shooting. Soldiers were allowed to go shooting only in parties of three, together with a native interpreter unless one of the three could speak the local language satisfactorily. A native cart and driver would be hired to carry the sportsmen and their equipment, and leave was freely granted from Wednesday afternoon until

Sunday night. Soldiers were strictly enjoined to respect local customs and keep away from villages and shrines. There was, however, sufficient uncultivated land in the vicinity of most stations to make it unnecessary to trespass, and there was plenty of small game for the marksman to shoot. Peacocks and monkeys were held to be sacred by Hindus, and troops were warned strictly against shooting them.

Organised team games became prominent in the later part of the nineteenth century. Many working-class boys, from whom the bulk of the Army was recruited, had neither the necessary equipment nor the playing space in industrial towns to take part in team games, and the Army gave them their first real introduction to football, hockey, and cricket. These games, in which the officers joined, were used, as in schools, to foster the spirit of competition and *esprit de corps*, as well as to encourage physical fitness.

Amateur theatricals and smoking concerts, the latter often containing songs and sketches full of topical and personal allusions, were even more popular in India than the United Kingdom, for there were few other distractions.

Non-commissioned officers and a proportion of the private soldiers could take their families to India at official expense, but single men had even fewer opportunities than their officers for respectable female companionship from members of their own social class. There was always the temptation to visit local brothels. A proposal that these should be regularly licensed and inspected was abandoned, when ladies of a moral vigilance society, vulgarly known as the 'Shrieking Sisterhood', appeared before Lord Roberts, arguing that official recognition of prostitution was the next thing to official condonation. The result was that at any one time, about 10 per cent of the British Army in India was in hospital suffering from venereal disease.

In 1901, just as Fraser was retiring, another soldier was about to enlist. This was Frank Richards, whose memoirs *Old Soldier Sahib* were to be published three years before Fraser's. His experiences and conditions of service were much the same as Fraser's, and both, like most old soldiers, are able to tell, and embroider, a good story. Richards was in a Welsh infantry regiment. His preoccupations and pastimes were much the same – girls, shooting, amateur concerts and the like. He too found the rations good – a pound of meat daily for each man, compared to three-quarters of a pound in the United Kingdom. Food was so cheap that a man could buy whatever took his fancy, if official rations did not suit his palate. The locally purchased beef, he thought, was tough. In camp native string cots or 'charpoys' could be bought, which kept the sleeper, his rugs and blankets off the ground, where everything might be eaten by termites or 'white ants'. Kit bags, if laced tightly, were, however, termite-proof.

Race relations

Richards was, just, a twentieth-century soldier and he began to encounter twentieth-century problems, one of which was the rise of Indian nationalism. He noted that Lord Curzon, the Governor-General, had set his face against the ill-treatment of Indians by British troops, and gives a completely false account of a rape incident which led to the 9th Lancers being disciplined. His version suggests that there was a little horse-play in a bazaar after some natives had treated the Lancers incorrectly. Because of Curzon's unreasonable attitude, it was now unwise to beat natives about the head, and instead blows had to be aimed at the body where they would be less apparent if any complaint were made later. He and the other young soldiers were regaled by an old European called 'the Bacon-Wallah' with tales of the good old days. The Bacon-Wallah had been a fusilier in one of the Company's European Regiments, and rather than transfer to the British line, had taken his discharge, settled in the country, and made a living supplying ham and bacon to the British troops. He felt the Army had gone soft. There was not enough looting or drinking done by his degenerate successors, in his opinion. The natives were becoming disrespectful. In his young day any native approaching an English soldier without salaaming three times would have been kicked off the veranda, he said. Richards claims that the Irish regiments were among the most conspicuous in terrorising and brutalising the local people. The Connaught Rangers in particular drew the colour line, and they drew it thick and firm. Their reaction when a local contractor told them that Lord Curzon had said Indians and British were brothers was to disprove this assertion by wrecking his establishment. The disbanding of this regiment (one battalion of which mutinied in India in 1921 because the Irish Special Constabulary were using on their own families at home the same techniques that they had themselves long practised on people in India), brought relief to camp followers and tradesmen throughout the sub-continent.

Assaults by British soldiers on their Indian fellow subjects were officially discountenanced, and were regarded as serious offences if proved. However, soldiers were often, in the heat of the moment, much too free with their fists, and boots, on the persons of the Indian servants and followers with whom they came in contact. The fact that in most cases the followers had neither the physical strength nor the daring to resist made this an even more unattractive trait, virtually amounting to sheer bullying. Many of these people suffered from enlarged spleens or other illnesses in which violence even of a minimal nature could set up complications leading to death. As the official 1935 handbook put it:

> Remember that if once the case gets into the law courts, there will probably be a lot of false evidence and exaggeration from witnesses on the other side, and it will almost certainly go hard with you. Though you may have only meant to administer a mild hiding, there can be no doubt that you are responsible for the damage done and you will have to suffer in consequence.

The same handbook also deals with the question of how soldiers should behave when travelling on the public railways. It admitted that the average Indian traveller probably no more desired to share his journey with a British soldier than vice versa. Most Indians were thought to be reasonably courteous if treated in the same way. Some, however, were prone to indulge in habits offensive to westerners, spitting betel juice on the floor being a particular cause of offence. On the other hand, the soldier was told he had no right to turn Indians out of a compartment so that he could travel alone. 'There is no getting over the fact that, so long as an Indian has paid for his ticket and behaves himself, he has as much right as you have in a carriage.'

In such circumstances, it is not surprising that Indian public opinion drew a firm line between the *sāḥib-lōg*, or 'the gentlemen' (the European officers and people of similar social standing) and the *goṛa-lōg*, or 'whitey', in particular the ordinary soldiers. The citizens of military garrison towns in the United Kingdom do not always relish the presence of soldiers in their midst. How much more had Indian citizens reason to fear and dislike the uncouth representatives of alien domination?

Relations between the British and Indian soldiers tended to be remote. With the exception of the extrovert Gurkhas, most Indian troops tended to be rather reserved in their dealings with the British. Although after the Mutiny each brigade had one British and two or three Indian units, there was little mixing. British troops could not be employed with Indians since Indian officers had no authority over British soldiers. This could lead to anomalies. For instance, an Indian officer could order his troops to fire on a rioting mob. A British sergeant, except in self-defence, could not. Yet the Indian officer could not order the sergeant to fire. Sita Ram noted the changed atmosphere between British and Indian troops after the Mutiny. Previously they had been on good terms, he said, the sepoys had done the English soldiers' guard duties in the heat, and stood sentry over their rum casks. But the new regiments lacked any understanding of Indian ways. 'The 17th Foot always called us *bhai* [brother – a polite term of address among equals]. The 16th Lancers never walked near our cooking pots or polluted our food, and we served with them for years.'

Yet how could the British forget the Mutiny, when every Sunday they marched to church taking their loaded rifles with them, to ensure that never again would there be a repetition of that fateful day of the Mutiny outbreak at Mirath in 1857, when the British garrison was unarmed, on church parade? How could they adopt racial attitudes other than the most unenlightened, when the only Indians with whom they really came into contact were low-class servants and cringing menials, or the type of sharp dealer that is attracted to military encampments in all ages and communities? Even the very words that the soldiers put to the urgent notes of their bugle call for the 'General Alarm', 'There's a

nigger on the wall, There's a nigger on the wall, There's a nigger on the wall', suggested the unfriendly circumstances in which they were most likely to meet members of the local population.

For soldiers serving on frontier stations, life was made additionally inconvenient by the activities of daring and expert thieves, known in army jargon as 'loose-wallahs'. The prime object of their depredations were the modern rifles and ammunition of the troops. Soldiers whose rifles were stolen were severely fined, and it eventually became a matter of unit or sub-unit pride not to lose a rifle. The most devious plans were resorted to. Rifles were chained together. The working parts were concealed and stored separately. Men slept with their rifles strapped to them. Some even buried them in the ground on which they slept. The 'loose-wallahs' proved able to surmount many of these precautions, and a large quantity of modern weapons did find its way into the hands of the tribes. There they were highly prized as status symbols, tangible trophies of daring, and an efficient way of killing enemies either in war against the British or in blood-feuds among themselves. Numbers of British soldiers kept dogs as pets, despite the risk of rabies. These animals could be 'neutralised' as defences by the loose-wallahs greasing themselves with leopard or cheetah fat. The leopard is well known to consider a dog a toothsome morsel, and Richards asserts that he saw with his own eyes how the scent of leopard would reduce the most inquisitive terrier to a bundle of fear, snuggling up as close to its master as possible, too frightened even to whimper lest it give its position away. Certainly it was common for thieves to grease themselves all over, making it more difficult for anyone to keep hold of them if apprehended. Murders by robbers in the course of such thefts were by no means uncommon.

The European soldier in battle

Whatever the shortcomings of the European soldier in peace, in war he was formidable. There are many instances of the courage and toughness of British troops in Indian battles. There were, too, some incidents when they hung back, or even turned from a particularly tough enemy, such as occurred more than once during the campaign against the Sikhs and Gurkhas. Nevertheless, in by far the majority of cases, it was the same story of relatively small numbers of European soldiers, supported by their sepoy comrades, defeating far greater numbers of equally brave and pugnacious foemen.

One case, although not a victory but a defeat, is a good example of the spirit such soldiers could show in adversity. The 1st Troop of the 1st Brigade of the Bengal Horse Artillery was the senior unit of the East India Company's army. They were the élite of all the Company's troops, and made much of their title, 'the First of the First'; the sergeant of the Number One gun of the battery claimed to command 'the Right Gun of India'. They formed part of the British garrison which was forced through incompetent generalship to retreat from Kabul through the snow-covered mountains of Afghanistan in January 1842.

The retreat was a disaster from the start. Afghan warriors closed in, cutting off stragglers, and rushing in on any small bodies detached from the main column. By the second day the column was beginning to disintegrate, and the Afghans grew bolder in their attacks. At the pass of Khurd Kabul an entire gun detachment of the troop perished to a man rather than lose their gun. But one by one, after the starving and exhausted teams failed to pull them through the deep snow, the remaining guns were spiked and abandoned. The troop, two officers and ninety-seven men strong at the beginning of the retreat, had now lost half its strength. The survivors carried on, turned their gun teams into troop horses, and served as dragoons. On several occasions they charged and drove off superior bodies of mounted Afghans, who gave them the sobriquet 'the Red Men' from the long red horsehair manes flowing from their Roman-pattern helmets. Gradually their numbers dwindled further. At Jagdalak most of the men and all the horses were lost, and the last two men joined the remnants of the 44th (Essex) Regiment and fought on as infantry. They were with the 44th when it made its last stand at Gandamak, leaving only one man, Dr Bryden, to struggle in to the British garrison at Jalalabad as the sole survivor of an army.

Kipling's India

Of fictional accounts of the British soldiers' life in India, the first and the best are to be found in the early works of Rudyard Kipling, the writer whose name is virtually inseparable from the image of the British *Rāj*. One of his earliest books, *Soldiers Three* (1890), introduces the characters of three privates of a British line regiment, whose stories, told as if in their own words, are full of well-observed details of life in camp, barracks, or battle.

These are based on talks he had with men of the 31st East Surrey Regiment (a south London unit), and of the Northumberland Fusiliers, in the garrison of Mian Mir (where he met our old friend, Sergeant Fraser). Several other volumes published in the 1890s and the novel *Kim* (1901) also contain passages describing ordinary army life. They reveal the way soldiers thought or behaved, as human beings, rather than the pipe-clayed automata or cardboard heroes of military historians and popular legend.

Charles Carrington, the biographer, writes, in *Rudyard Kipling, His Life and Work*:

> Search English literature and you will find no treatment of the English soldier on any adequate scale between Shakespeare and Kipling. He who wishes to know how British soldiers fight, how officers and men regard one another, how they talk the night before the battle, will seek the information in *King Henry V* or in *Barrack-Room Ballads*, for it is to be found nowhere else in our English classics.

Kipling's attitudes often mirror those of the Anglo-Indian society to which he belonged. His admiration for soldiers and the military virtues accords with the views of men whose jobs, whose position, whose very lives and those of their wives and families were safeguarded by the presence of British soldiers and their Indian auxiliaries. By contrast, most citizens of the United Kingdom then, as now, regarded military officers as elegant dolts and their men as brutish louts. Kipling also had a love for the hierarchical structure of government, for the grandeur of official titles (like most Anglo-Indians, he preferred to use the term 'Viceroy' rather than 'Governor-General'), for the services in which each had his place in the order of things. Kipling's India contained no British aristocrats. Rarely did British noblemen accept posts in India, and then only in the highest offices for terms of strictly limited duration. Lacking a hierarchy of birth, the Anglo-Indians evolved one of office. Government service, generally despised at home, was regarded in India as the most honourable of professions, and so, too, by Kipling. The hero of *Kim*, using his knowledge of the people of India to uphold the British *Rāj*, finds what is evidently intended to be an estimable career by joining an Indian version of the Secret Police!

Because of Kipling's genius in transmuting anecdotes or tales he heard into stories that have the ring of truth about them, many of his statements, written as fiction, could be taken by his readers as fact. Kipling's characteristic trick of stepping outside his narrative, to clinch his story with some apparent assertion of fact, is really part of his technique of telling a tale. But, as in the novels of another authoritative writer on India, John Masters, while the stories are set in realistic situations, and the details of the backgrounds are authentic, the characters, their actions and words are fictitious.

An illustration is to be found in the last words of Kipling's *The Jungle Book*. These conclude a story ('Her Majesty's Servants') which is not about the wild beasts of the jungle, but about the camp beasts used by the military: a troop-horse, gun-bullocks, a draught elephant, ordnance mules, and a pack-camel. It is set at the great military review at Rawalpindi, in March 1885, which Kipling covered as a reporter for his newspaper. This review was one of the many ceremonial occasions organised in connection with the state visit to India of Abdul al-Rahman, Amir of Afghanistan. In the story, Kipling, and his fox-terrier, eavesdrop at night on the conversation of the different beasts, talking about their men, and their own reaction to the military life. The tale ends with a set-piece description of the review the following morning, a piece of writing that conveys a most vivid impression of the scene. At the end, Kipling makes an old Afghan chief of the Amir's bodyguard ask questions of a native officer – how was this done, and were the beasts as wise as the men?

> 'They obey as the men do. Mule, horse, elephant, or bullock, he obeys his driver, and the driver his serjeant, and the serjeant his lieutenant and the lieutenant his captain, and the captain his major, and the major his colonel, and the colonel his brigadier commanding three regiments, and the brigadier his general, who obeys the Viceroy, who is the servant of the Empress. Thus it is done.'
>
> 'Would it were so in Afghanistan!' said the chief; 'for there we obey only our own wills.'
>
> 'And for that reason,' said the native officer, twirling his moustache, 'your Amir whom you do not obey must come here and take orders from our Viceroy.'

Thus writes Kipling, the teller of tales for children, the advocate of law, order, and Empire. The historian, with the task only of saying what really happened, must record that it was only four years before this review that this carefully organised Indian Army had failed, most signally, to subdue these wild Afghans; that there was no 'must' about Abdul al-Rahman's visit to India, but that on the contrary it was a considerable triumph of diplomacy to persuade him to come; and that so far from his 'taking orders', he went back to his own country the recipient of what, in the jargon of a later age, is called 'military aid without strings'.

Elsewhere Kipling was to write that Truth is a naked lady, and it behoves a gentleman to give her a print petticoat or turn his face to the wall and vow that he did not see. But that, as he himself would say, is another story.

9

Operations: Southern Afghanistan, July–September 1880

This campaign gives a fair idea of the capability and performance of the Indian Army against another Asiatic army, operating beyond the North-West Frontier of India.

During the 1870s, British statesmen were increasingly alarmed at the continued expansion of Russian power in Central Asia. They feared that, at a period of tension in Europe, the Russians would distract the British there by threatening India and stirring up trouble among the border tribes. The Prime Minister, Benjamin Disraeli (Lord Beaconsfield) sent Lord Lytton to India as Governor-General, with instructions to establish British influence over Afghanistan, and to forestall any Russian advance in that direction.

Sher Ali, the Amir of Afghanistan, resisted British demands to send an envoy to his court and to establish a Resident or permanent ambassador there. He wished only to be independent, he said, and would entertain neither a Russian nor a British presence in his country. He would not be able to guarantee the safety of a British Resident. His people remembered the Afghan War of 1839–42, when a popular rising had led to the murder of the last British Resident, and the destruction of his army. Lytton insisted that an envoy be accepted. After some months delay the proposed British envoy, General Sir Neville Chamberlain (perhaps better known to posterity as the inventor of the game of snooker), attempted to enter Afghanistan with his escort, but was turned back by Afghan frontier troops. This was taken as an insult to the British. Lytton declared war, and ordered a full-scale invasion.

The first part of the war was successful. Large numbers of British and Indian troops invaded Afghanistan. The Afghan army was defeated and forced back. The Amir Sher Ali fled, and the British signed a peace treaty with his successor, Amir Yakub Khan, who accepted the appointment of a Resident at Kabul, the Afghan capital. The troops began to withdraw to India. Stores and transport animals were auctioned off, and the war seemed to have reached a satisfactory conclusion.

In September 1879 the British received bad news. Soldiers of the Afghan regular army stationed in Kabul had mutinied. The Amir had done nothing to restrain them, and, supported by the city mob, they had attacked the British Residency. The British Resident and all his escort had been massacred.

Back came the British troops to avenge the murders, only to find themselves with another major campaign on their hands. The Afghans fiercely resisted the invaders of their homeland. To patriotic fervour was added religious fanaticism, as these zealous Muslims took up arms against an army led by Christians and containing a large number of Hindus and Sikhs, idolators and unbelievers. Kabul was reoccupied after severe fighting. Amir Yakub Khan was deposed and sent under arrest to India, but still disturbances continued.

In the United Kingdom the campaign became unpopular with the electors. Just as the religious leaders in Afghanistan preached holy war against the invaders, so did Mr

Gladstone, the leader of the Liberal party, fulminate against those who ordered the invasion. The sanctity of life in Afghan hill villages became an issue in the General Election of 1880. Public opinion was shocked at the spectacle of British armed might being used to bully a small Asian state. It was horrified to hear of hamlets burnt and their inhabitants slain in defence of their homes. It was also unhappy at the failure of British troops to win a decisive victory, despite the technological superiority they enjoyed over the Asian peasants in arms against them. Gladstone's Liberal party swept the country. Disraeli and the Conservatives were turned out of office. Lytton resigned and was replaced by a new Governor-General, Lord Ripon, whose orders were to end the war and withdraw the troops to India as quickly as possible.

By July 1880, eastern Afghanistan was quiet. The British agreed to recognise, as Amir, Abdul al-Rahman, a member of the Afghan royal house and a good general. He was able to establish a stable and relatively popular regime, independent of British support, and the British officers at Kabul began to plan their army's withdrawal to India.

But, at the same time, a serious crisis developed in southern Afghanistan. The Wali, or Governor, of the city and province of Kandahar was a British ally. He was also inefficient and unpopular. Ayub Khan, another Afghan prince, claimed the right to rule Kandahar. He was an able commander, and had an army which included many of the old regular regiments originally raised by Amir Sher Ali. He was supported by the local people with both supplies and men, and it was clear that his challenge had to be taken seriously.

At the end of July, Ayub left his base at Herat, in western Afghanistan, at the head of a strong force. His regular infantry was organised in three brigades, each of three battalions about 500 strong. Some units were armed with muzzle-loading Enfield rifles, others with breech-loading Sniders of the same pattern as those carried by the Indian troops. His cavalry was divided into three regular regiments, each 300 strong, and about 2,000 irregular horse. His artillery amounted to thirty guns. These were divided into four field batteries, (including one of six 12-pounder Armstrong rifled breech-loaders), with a mountain battery of 3-pounder guns carried on pack-ponies. Each battery was manned by about a hundred gunners, artificers, and drivers. As he marched he was joined by other regular units and numerous religious fanatics or Ghazis (*G͟hāzī*).

The battle of Maiwand

At this time the British garrison in Kandahar, under Major-General Primrose, an elderly veteran of the Indian Staff Corps, totalled about 4,600 men, British and Indian, of the Bombay Army. They were made up of three cavalry regiments, three batteries of artillery, two European battalions and three Indian battalions of infantry, with a company of sappers and miners.

As Ayub advanced, the Wali's power crumbled. Men flocked to Ayub's camp, while large numbers left Kandahar with their families and belongings for fear of the coming disturbances. The price of food soared, and hoarding began. In an attempt to restore the Wali's diminishing authority, a force under the ageing Brigadier-General G. R. S. Burrows of the Bombay Army was sent out from Kandahar, to join the troops of the Wali's army and prevent Ayub crossing the River Helmand. Unfortunately the Wali's men were so impressed by Ayub's success that some deserted to his side and the rest mutinied and disbanded themselves. Burrows' cavalry pursued the Wali's artillery, which was heading for Ayub's camp, and were successful in capturing its guns. These were all smooth-bores, four 6-pounder field guns and two 12-pounder howitzers. Ammunition supplies were limited, but on the suggestion of Captain Slade, battery captain of the Royal Horse Artillery in Burrows' force, they were added to the British order of battle. Slade became battery commander. The three 'divisional officers', who each commanded a division of two guns, were provided by a subaltern of the 66th Foot and two Royal Artillery officers who were doing duty with the Ordnance and Transport staff of the column. Forty-two men from the 66th, who by a curious coincidence had been given some training in gun drill on a previous occasion, became the gunners of the battery, with a few soldiers from the Royal Horse Artillery.

Burrows' force now consisted of a strong brigade. There were two cavalry regiments, the 3rd Bombay Light Cavalry and the 3rd Scinde Horse, the former with 6 British officers and nearly 300 men, the latter with 5 officers and 250 men. The artillery consisted of E–B Battery of the Royal Horse Artillery (six 9-pounder light field guns) with a posted strength of 5 officers and 189 men, with the smooth-bore battery recaptured from the Wali's army.

The Infantry was made up of the 66th Foot, with 19 officers and about 500 men, the 1st Bombay Infantry (Grenadiers), with 7 British officers and 641 men, and the 30th Bombay Native Infantry (Jacob's Rifles), with 8 officers and 617 men. Engineer support was provided by a half-company of Bombay Sappers and Miners.

Little was known of the numbers and proximity of the enemy. Cavalry patrols were sent out in the usual manner of troops in the near presence of an enemy. Little was discovered of any value, and some patrols were intercepted and cut down by enemy horsemen. Spies, paid by Colonel St John, the political officer responsible for the conduct of British relations with the Wali and his erstwhile subjects, reported that Ayub was moving on the village of Maiwand, about forty-five miles west of Kandahar. Burrows decided to anticipate Ayub by marching on Maiwand and so placing himself between Ayub and his obvious objective, Kandahar itself.

At 4.30 am on 27 July 1880, his column fell in to advance on Maiwand, which was some six miles away to the north-east. All stores, ordnance and commissariat supplies accompanied the column, as it was unsafe to leave them without a strong guard, and the force was too weak to afford any large detachment.

The column finally moved off at about 7 am. The rate of march was slow, with several halts to allow the baggage train to close up. At about 10 am another of St John's spies reported that Ayub was occupying Maiwand in force. Burrows decided to go on, and sent forward cavalry patrols to discover what was happening. However, the fact that Major Hogg, Brigade-Major of the Cavalry, went some 800 yards beyond the furthest outposts and was back with Burrows after an absence of ten minutes indicates how short was the distance that their reconnaissance covered. It was clear, nevertheless, that large numbers of the enemy were actually marching across the British front from Maiwand towards Kandahar, and Burrows decided to attack. The terrain was an arid, treeless, undulating plain, without cover of any kind except for ravines about fifteen feet deep to the British right and rear.

The guns and cavalry went forward until halted by large numbers of enemy horse.

The two batteries came into action, with the smoothbores in line on the left, and the RHA in echelon of divisions on the right. They were protected by the cavalry on their flanks and rear. The infantry formed a line in their rear, the Grenadiers on the left, half the Rifles in the centre, the 66th on the right. The rest of the Rifles, with the Sappers, were some 200 yards in the rear as supports. Behind these the baggage and field hospital, with an escort of a company of the 66th, one of the Grenadiers, and one of the Rifles, were still being collected and formed into some sort of order for local defence.

Finding that their cavalry had halted the British advance, the Afghans brought up their guns and regular infantry. The British infantry then came up to form a line with their guns, and the British position, at about 11 am, was as follows. On the right flank were the 66th Foot and half of Jacob's Rifles. Then came a division of horse artillery under Lieutenant Osborne; two companies of Jacob's Rifles; Lieutenant Fowell's division of horse artillery; the half-company of Sappers, fighting as infantry; Lieutenant Maclaine's division of horse artillery; the Bombay Grenadiers; the smooth-bore battery, and on the left flank two more companies of Jacob's Rifles. The cavalry concentrated on the left flank where large numbers of enemy irregular horse and foot were continually threatening to outflank the British position and take the baggage lines.

The Afghans, with the advantage of superior numbers, rapidly built up their firing line, and extended it in an arc across the British front, so that they could concentrate their fire inwards on the much smaller British line. The Afghans had been just as surprised as their enemies to find a large body of men drawn up in their vicinity, and a conventional 'encounter battle' took place. Ayub Khan had increased his forces since leaving Herat, and within about an hour of the beginning of the battle he was able to concentrate them in overwhelming force. In addition to his thirty well-served guns (including the six Armstrong guns which threw a heavier shell than anything Burrows' gunners could pit against them) and his nine regular battalions, he could field four regular cavalry regiments (three of his own army and one of the Wali's mutineers), a thousand of the Wali's infantry, fifteen hundred irregular horse, and several thousand ghazis and other tribal warriors.

A fire fight developed, in which the superior numbers of the Afghans, especially in artillery, began steadily to tell. The Indian cavalry on the left flank suffered badly. Their elderly brigadier, Nuttall, feared to withdraw them in case the Afghan irregular cavalry attacked the British flank and rear, and so they remained, in a treeless plain, unable to take cover from a cruelly accurate fire. On the British right flank the Afghan gunners used the ravine, or *nala*, which ran across their front and round the side of the British, to bring up their 9-pounder guns, first to 700 yards, and then to 500. Finally, under the cover of these batteries, two 3-pounder mountain guns were pushed up to about a quarter of a mile from the British line.

Hitherto the casualties had been mainly among the troop horses and the exposed gun teams of the cavalry and artillery. Now the infantry began to suffer, both from the guns and from the regular Afghan infantry with which regular volleys were being exchanged. Swarms of ghazis pushed on down the ravine towards Khig, a village in the British left rear, and the British line, endeavouring to prevent itself being outflanked on both sides, bent back until it assumed the shape first of a bow, then a semi-circle, and finally a horseshoe.

Slade's smooth-bore battery had gone into action with two extra disadvantages. Firstly, no officer had been appointed to the post of battery captain (or second-in-command) with the duties in action of taking charge of the wagon lines and ensuring the flow of ammunition from the reserve to the gun line. Secondly, the battery, in fact, had scarcely any wagon lines at all, because when its guns had been captured from the Wali's men the drivers had cut the traces of the ammunition wagon teams and galloped away on them. The wagons had then been impossible to move and had been destroyed with their precious ammunition, to prevent their falling into Ayub's hands. Now, following a prolonged artillery duel, the smoothbores ran short of ammunition, and as there were no wagons with the guns and no battery captain to deal with the re-supply problem, Slade ordered his men to limber up and fall back to the ordnance park to replenish their diminished stocks with the little that was kept there. He could not send the limbers alone, because this would have left the guns immobile in the face of an advancing enemy. Slade then rejoined E–B Battery, of which he was normally second-in-command, and was ordered by the battery commander Major Blackwood to take command. Blackwood had been badly wounded in the thigh by a shell splinter, and had to fall back to have the wound attended to. He was in fact unable to mount his horse again and take command of his battery, so went across to the 66th Foot, where he helped their commander, Colonel Galbraith, in giving to the men the correct range of their targets.

It was now about 2 pm and the British position was growing precarious. Large numbers of Afghan irregulars had moved round the flanks of the firing line, and the baggage escort, composed of three companies, one each from the 66th, the Grenadiers, and Jacob's Rifles, was now fully committed against these. On the left flank the two isolated companies of Jacob's Rifles had, like the cavalry, suffered severely from the Afghan artillery. They had lost their only British officer, Lieutenant Cole, early in the day, and all three of the Indian officers with them had been badly wounded.

At this point, seeing the withdrawal of the smoothbore battery, a crowd of ghazis rushed on the British line. These fanatics were a most formidable enemy in close combat. Most of them were armed with long Khyber knives, which resemble those commonly to be seen in butchers' shops. In the grip of religious frenzy, and convinced that they would achieve immortality if slain in battle against the infidel, many actually foamed at the mouth as they charged. Their wild appearance was too much for the two companies of Jacob's Rifles, who abandoned their position and fled.

The Grenadiers were then exposed to the full brunt of this attack from their left flank. Prior to this, they had been lying down to avoid the artillery. Standing up, they tried to form square, but the small number of officers with them were unable to do more than form a sort of 'V' formation. The regiment became hopelessly disordered, and finally broke and fled. The panic of the fleeing Riflemen spread to the Grenadiers, many of whom stood irresolute until cut down by the ghazis in amongst them.

The main body of Jacob's Rifles then gave way and crowded to their left upon the 66th Foot, which alone retained its formation. Firing steadily, the 66th fell back upon Khig, amid a mixed crowd of fleeing sepoys and exultant ghazis.

The artillery, left unsupported, had to fall back, but lost Lieutenant Maclaine's two guns on the way, the gunners and drivers only just getting away in time. This was a horse artillery battery, and so the gunners had their own horses, and fought their way out, with the drivers and limbers. Lieutenant Osborne was killed, and Lieutenant Fowell had already been sent to the rear badly wounded, so Slade and Maclaine had now to command the four remaining guns alone. The smoothbores came into action to cover the retreat, and then, having fired away the last of their ammunition, fell back to the baggage train where the rearguard had acted as a rallying-point for some of the men from the broken firing line.

Burrows ordered Nuttall's cavalry to charge the ghazis, to try to give the infantry a chance to reform. However, the troopers, now reduced to about 130, could not be induced to charge, nor did their officers form them into a proper line. The cavalry instead cut its way out to the rearguard, where it was eventually brought back into an orderly formation.

The leading companies of the 66th passed through Khig and made a stand in a walled garden beyond. They were then surrounded by almost the whole Afghan army which was following up the British retreat. The remnants of the regiment, about a hundred strong, closed in upon their colours, men falling one by one until at last only about a dozen were left. These valiant soldiers then rushed out upon their foes, and died fighting. Among the officers killed here were Colonel Galbraith, two captains, and five subalterns of the 66th, and Blackwood of the Artillery, together with Lieutenant Hinde, adjutant of the Grenadiers, and Lieutenant Rice Henn of the Sappers, who with some of their men had joined the 66th as formed bodies and fought with them to the last.

The rear companies of the 66th fell back to their left and joined the baggage, which was now streaming away from the scene of the fighting. Baggage animals, stores, wagons, were all abandoned in what had become a rout. The last stand of the 66th had shown the Afghans that the British could still fight hard if cornered, and the prospect of looting the abandoned stores was more attractive than further fighting, so there was no close pursuit. Indeed the Afghan army had suffered almost as much as the British from the great heat of the day, from thirst, and from the fatigue of marching and fighting for nearly twelve hours. The artillery alone of the British force retained any military formation, and the small bodies of Afghan cavalry that did follow up the rout were kept at a distance by Slade's guns. The cavalry who should have protected the rear refused to do so and in fact led the retreat, gradually drawing further and further ahead of the main body.

Dreadful scenes followed as the retreat continued through the night. Men and horses collapsed from exhaustion and lack of food and water. Maclaine was taken captive trying to find a water point. When one at last was reached a wild rush was made for it, and men and beasts struggled to reach it during a halt of some ninety minutes. Many had not yet had a drink when the report of another Afghan advance led to the retreat beginning again. Many men were left behind, and stragglers were continually being picked off by the local inhabitants who had risen against the defeated invaders. One by one the smooth-bore guns had to be abandoned as their exhausted teams could pull them no further, but the limbers were brought on, crowded with wounded men, for all knew that anyone left behind would be tortured and murdered.

At Kandahar, Primrose, the garrison commander, was awakened at about two in the morning with the news of the disaster. Confirmation of Burrows' defeat and that he was retreating towards Kandahar came an hour or so later, and the sound of shooting from the adjoining villages informed the general that his information was shared by the locals and that the countryside was up in arms against the British.

A force was at once ordered out to cover the retreat, and at about 5.30 am a column under Brigadier-General Brooke left Kandahar. These men, one troop of the Poona Horse, two guns of C-2 (Field) Battery, one company of the 7th Royal Fusiliers and another of the 28th Bombay Infantry, met the remnants of Burrows' force about four hours later. Their appearance gave the exhausted survivors a chance to halt and allow some of the stragglers to come up. Then the remnants of Burrows' column set off for Kandahar, while Brooke's troops fought off several attacks on their front, rear, and both flanks.

The whole force returned to Kandahar at about 2 pm. The Maiwand column had now marched something like fifty miles and had been on the move for thirty-three hours. Out of just under 2,500 soldiers engaged in the battle, 21 British officers, nearly 300 British soldiers, and 650 Indian troops were dead, and 168 all ranks wounded. A further 786 followers were killed or missing. The cavalry and artillery had had 201 horses killed, with 68 wounded. The transport animals lost totalled 1,676 camels, 355 pack-ponies, 24 mules, 79 bullocks and 291 donkeys. Over 1,000 rifles were lost, with many hundreds of swords and bayonets. No smoothbore ammunition was lost, all 487 rounds having been fired away. The horse artillery fired 1,473 rounds, but a further 448 were left behind with the baggage train, with 278,200 rounds of small-arms ammunition.

All the sick and wounded men were brought in from the outlying cantonments, and at dusk, there being no sign of any more stragglers coming in, and with increasing Afghan activity in the neighbouring areas, the last British picquets were withdrawn, and the gates of the city wall closed. Primrose then set about organising his defences for a siege.

The siege of Kandahar

The garrison totalled just over 5,000 all ranks, with fifteen guns, organised as follows:

		Officers	*Men*
Cavalry	Poona Horse	5	224
	3rd Scinde Horse	5	367
	3rd Bombay Light Cavalry	6	383
Artillery	E–B Battery, Royal Horse Artillery (with four 9-pounder guns)	4	134
	C–2 (Field) Battery RA (less one division) with four 9-pounder guns	5	115
	5–11 (Garrison) Battery RA with four 40-pounder guns and two 8-inch mortars	4	90
	The remaining smoothbore 6-pounder was also available		
Infantry	7th Royal Fusiliers	24	620
	66th Foot	12	241
	1st Bombay Grenadiers	6	345
	4th Bombay Native Infantry	7	554
	19th Bombay Native Infantry	6	613
	28th Bombay Native Infantry	7	707
	30th Bombay Native Infantry (Jacob's Rifles)	5	396
Engineers	No 2 Company Bombay Sappers and Miners	1	66

There were also about 1,000 cavalry horses, 200 artillery horses, 372 draught oxen for the heavy battery, another 135 transport bullocks, 602 ponies, 68 mules, and 1,021 camels.

Kandahar city was roughly a rectangle, with two short sides – the south about 1,300 yards long, the north about 1,200 yards long – and two long sides – the west about 2,000 yards long and the east about 1,700 yards long. The average height of the mud walls was some 30 feet, and the width about 15 feet. There was a good ditch some 18 feet deep on the north and west fronts but none on the south. The walls were dotted at intervals with a total of forty-nine bastions, much dilapidated, but still able to bring flanking fire to bear on any storming parties. There was a gate in the centre of each short wall and there were two gates in each long wall, each with bastions on either side of it. The parapet running along the top of the main walls was some eight feet high, with loopholes and embrasures for musketry.

Standing high above the rest of the town just inside the north wall was the citadel. However, this would have made an inadequate refuge, as its walls were weak and its water supply poor. The aim of the defence was therefore to secure and hold the city perimeter wall at all costs.

Working parties were at once set to strengthening the defences. The numerous canals and underground passages passing through the walls were filled up and mined with fougasses. These, with the barbed wire and sandbags placed on the parapet, deterred the ghazis from attempting to rush the walls. All gateways were protected by thickets of thorn-bush, and the gates themselves strengthened with plates of sheet-iron. All trees, buildings, and walls around the city were levelled to create an *enceinte*, an open space around the perimeter that would afford no cover to the enemy. A barbed-wire entanglement was placed at the foot of the whole wall. Telegraph lines were run between all the gates and the headquarters in the citadel, and, as the last word in up-to-date technology, a telephone line was run from the citadel to the north-west bastion. Visual signal stations were installed at each corner of the defences. The infantry was distributed along the walls ready to beat off any attack. A special squad of sharpshooters was formed to pick off enemy snipers. One heavy gun and the two mortars were deployed in the citadel, the other three 40-pounders in three of the corner bastions. The 9-pounders and the remaining 6-pounder were distributed to cover the gates and the northern corners. All wells were located and allotted to various units. There were ample stocks of ammunition, and enough food and forage to supply the troops and cattle for several weeks.

The remaining source of weakness was the civil population of the town. Most were sympathetic to their fellow countrymen outside, and the possibility of their forming a 'fifth column' within the defences was very real. At the very least they represented a number of useless mouths consuming the available stocks of water and food, which, ample for an army, would not supply a whole city. Brigadier-General Brooke, a young and active British Army officer, advocated their expulsion. Colonel St John, the diplomat, strongly opposed him, pointing out that the British, unpopular already, would be even more hated if they turned out from their homes thousands of non-combatants, old people, women and little children. Primrose decided that military necessity obliged him to follow Brooke's plan, and on 29 and 30 July the whole Afghan population was evacuated. The arrival in Ayub's camp of thousands of refugees, carrying with them what few possessions they could, roused the Afghan army to a new pitch of anger against the invaders of their country.

During the first week of August, Ayub gradually brought up his army and tightened his grip round the city. All communications with the outside world were cut off. Messengers dispatched through the Afghan lines were intercepted, and killed in an unpleasant manner. A steady exchange of musketry began between the defenders and the besiegers, and two of Ayub's Armstrong guns began a sporadic bombardment.

As the siege went on into its second week it became clear that Ayub was settling down to a careful and thorough investment. The suburbs of the city were now occupied by him and were being fortified as points from which an attack would eventually be made. The position of the mass of his artillery was still unknown to the British, as only two out of the thirty guns had been unmasked.

On 15 August the defenders made a sortie to drive the Afghans back. The village of Deh Khwaja, about half a mile to the east of the city, was selected. The main body of the Afghan army was encamped on the west of the city, and their supports coming up to help the men in Deh Khwaja would be exposed to British gunfire and to the cavalry which was ordered out to cover the sortie. At first all went well. The infantry columns advanced steadily out from the city and occupied the village. The Afghan reinforcements were intercepted and driven back by the Poona Horse and 3rd Light Cavalry according to plan. Further Afghan attempts to advance were stopped by infantry fire from the flank guards, supported by artillery from the walls. The village was occupied and the Afghan garrison driven out.

Then, as the force began to return to Kandahar, the British made a fatal mistake. The cavalry and infantry supports were recalled before the main column had left the village. At once large numbers of ghazis and tribesmen rushed up from the south with the object of cutting off the British from the city. Firing from the shelter of walls and field boundaries, they inflicted severe casualties on the retreating column, which was crowded along a narrow road and unable to make an effective reply. The British suffered casualties of 108 killed and 120 wounded out of the 1,556 men engaged in the sortie. There were very heavy losses in officers, ten of whom were killed or mortally wounded. Brigadier-General Brooke, the column commander, was killed trying to save the life of one of the Royal Engineer officers. The Reverend Gordon, of the Church Missionary Society, who had accompanied the force as a chaplain, was also shot and killed while attending under heavy fire to the wounded and dying.

All the troops (mostly from the 7th Fusiliers, 19th and 28th Bombay Infantry) showed great steadiness throughout the fighting, and there was no repetition of the panic that had occurred at Maiwand.

The siege continued for another fourteen days, but no serious assault was made on the defences. Sniping and occasional shelling continued, and further entrenchments were made by the Afghans. Ayub, however, could not afford to throw men away in an attempt to storm the walls. He had to keep his army intact, not only to deal with the British who would come to revenge their earlier defeat, but also to deal with the forces of Amir Abdul al-Rahman, who would not be content to rule only half Afghanistan when the British had gone.

Meanwhile, in the outside world, the news that a British brigade had been wiped out at Maiwand and that the best part of the division was shut up behind the walls of Kandahar caused general consternation. In the United Kingdom troop reinforcements were ordered for India, and there was a public demand that Sir Garnet Wolseley, the 'trouble-shooter' of the late Victorian Empire, should be sent to take command. The senior naval officer at

Rangoon offered the services of a naval brigade. The Bombay Army gathered its forces to the south, and prepared a relieving column from the troops in Baluchistan under Major-General Phayre. At Kabul there was a black moment as the officers there wondered if their own position could be held if the rising spread northwards.

It became clear almost at once that Phayre would find it difficult to make a speedy march. The railway intended to run from Upper Sind to the Bolan Pass, through which traffic from India entered southern Afghanistan, was still under construction. He had insufficient animal transport for his needs, and a two-year drought meant there was little grain and fodder available for what he had. Moreover the Baluch tribes, seeing the British in difficulties, took advantage of the movement to resume their ancient occupation of raiding the country between Afghanistan and India. One such raid by about 2,000 men on the town of Kach, held by 300 men of the 16th Bombay Infantry, was driven off with severe losses to the tribesmen, but served to delay the relieving column even more.

The relieving column

The commander of the British troops at Kabul, Lieutenant-General Sir Donald Stewart, organised a flying column under Lieutenant-General Sir Frederick Roberts, who had in fact held a separate command prior to joining forces with Stewart's troops a few months previously. The original suggestion seems to have come from the Commander-in-Chief, India, General Haines, though Roberts later claimed the credit for it. Roberts was given the pick of the troops available. Many of the Kabul garrison were only too glad not to be selected, as they had been looking forward to returning to India after more than two years campaigning in a strange country, full of mountains and deserts, alternately too hot and too cold.

A strong division of all arms was assembled and styled the Kabul–Kandahar Field Force. The cavalry brigade was made up of the 9th Lancers and three Indian regiments, the 3rd Bengal Cavalry, the 3rd Punjab Cavalry (from the Punjab Frontier Force) and the Central India Horse, a unit which had been borrowed from the Government of India in the Foreign Department. The artillery consisted of three mountain batteries (6–8 and 11–9 Royal Artillery), each with six of the new 2·5-inch 'screw-guns', and No 2 Indian Mountain Battery with six of the older 7-pounder guns. No wheeled artillery was taken in case the column had to leave the roads and move across country.

There were three infantry brigades. The 1st was made up of the 92nd Highlanders, 23rd (Punjab) Bengal Native Infantry (Pioneers), 24th (Punjab) Bengal Native Infantry, and the 2nd Gurkha Regiment. The 2nd Brigade consisted of the 72nd Highlanders, with the 2nd and 3rd Sikh Infantry and the 5th Gurkha Regiment, all three from the Punjab Frontier Force. The 3rd Brigade was composed of the 2nd Battalion of the 60th Rifles, the 15th (Sikh) Bengal Native Infantry, the 25th (Punjab) Bengal Native Infantry, and the 4th Gurkha Regiment. The battalions were on average about 600 strong, the British with about 21 officers each and the Indian with about 8. There were no sappers, and the responsibility for field engineering was undertaken by the 23rd Pioneers. As a result of the Indian Mutiny, twenty years before, when so many Bengal Army officers had been killed, the average age of unit commanders of these regiments was much younger than those of the Bombay and Madras corps.

All transport was by pack-animal, and stores were reduced to a minimum. Each European officer was allowed one mule for his personal effects. British soldiers were allowed 24 lb of baggage per man, and each Indian soldier 20 lb. Thirty days' supply of tea, sugar, salt, and rum was carried for the British troops, with two days' supply of preserved meat and five of breadstuff. For the Indian troops five days' supply of their staple flour, *ātā*, was carried, thirty days' of *dāl* (pulses or lentils) and salt, and enough sheep for ten days. Two hundred gallons of limejuice also accompanied the column, with a small amount of preserved vegetables. Each infantryman carried 70 rounds of rifle ammunition in his pouches. A further 30 rounds per man were carried in the regimental reserves, and 100 rounds per man in the Ordnance Field Park. The batteries each had 540 rounds of gun ammunition with them, and there were a further 30 rounds per gun in the Ordnance Field Park. One mule was allotted to each troop or company to carry the arms of men falling sick. Quartermasters, adjutants, armourers, paymasters, farriers, saddlers and pioneers were all allotted their special share of transport. The field hospital had no wheeled ambulances but was accompanied by 2,192 bearers, 115 doolies, 286 ponies, 43 donkeys, 3 bullocks and 6 camels. The rest of the column's stores was carried on the backs of 2,740 pack-horses and ponies, 4,511 mules and 912 donkeys. There were 1,779 cavalry troop-horses, 450 artillery mules, and more than 200 chargers for the staff and field officers. The whole force, totalling 2,836 British all ranks, 7,151 Indian troops, and some 7,000 followers moved off on 8 August.

The march which followed is one of the most rapid in the annals of military history. In twenty days, with only one day's rest, it covered 280 miles. Communication by heliograph was established with the Kandahar garrison on 27 August, and contact was made with patrols from the beleaguered city on the same day.

From a military point of view the operation was by no means as daring as it was popularly hailed to be, or as was suggested in Roberts' memoirs. Although he was marching through a country whose inhabitants were of unknown disposition, without a base of operations or any lines of communication, in the direction of a victorious enemy force, in point of fact there was no reason why danger should have been anticipated. The local inhabitants, in the main, were supporters of Abdul al-Rahman, who was anxious that the British be given every assistance to speed them on their way to defeat his rival Ayub. The troops were able to obtain most supplies by local purchase from the villages through which they passed. Their numerical strength was such that no enemy they were likely to meet could outmatch them. Only Ayub's thirty well-served cannon were a threat, but these could not be pitted against Kandahar and its relievers simultaneously, nor could Ayub leave the city unwatched in case another sortie were to be made. Stewart himself had made the same march in the other direction the previous year and saw no danger in sending Roberts other than the risk of weakening his own force which at the same time was withdrawing from Kabul to India.

> Bobs [Roberts] is very eager to go [he wrote to his wife] and I dont wonder; though people say it is not the right thing to do. It is what we did last winter with a much smaller and less efficient force than the one I have now told off. I am giving Bobs nine regiments of Infantry while I only had seven; and he will have three European regiments of Infantry to my two and the 9th Lancers in addition. He will also have two Gurkha and two Sikh regiments; so his force in fighting power will be nearly

twice as strong as my Division, good though it was. Still it is only fair to give him everything and risk as little as possible (31st July).

... This is a grand thing for Bobs. If there is any fighting he can't help being successful, and his success must bring him great credit.

The real achievement of the march was the work done by the administrative staff and their supporting teams. Divisional and brigade staff officers arranged routes, allotted camping grounds, planned movement tables. Regimental adjutants and quartermasters and the battery captains supervised the myriad duties of their office, supervising the order of march of their units, seeing to the distribution of supplies, and arranging the administrative details of the move. Sergeant-majors and havildar-majors, with their clerks, storemen, and other underlings, dealt with the daily routine of controlling the loading, unloading, packing and unpacking of tents, bedding, cooking utensils and other domestic equipment. The transport officers toiled to keep the long trains of baggage animals on the move, or to see to their food and water and lines when halting. The commissaries went out to buy livestock, grain, and firewood from the surrounding country to feed the army and conserve the limited rations carried with them. The grain was supplemented locally and on only two occasions did the British soldier have to forego his western-style bread and accept the *capātīs* or unleavened flour cakes of the native troops. The men of the field bakeries and butcheries, the drivers, the bearers and other followers performed prodigies of industry keeping the army's supports on the move and making things easier for the men of the fighting line. Their efforts, indeed, were an essential part of the success-story of this column, since the principle 'no labour, no battle', which is one important to all fields of warfare, was very important indeed in the campaigns of nineteenth-century India.

All the members of the column had to endure their share of hardships on the march. Long stretches of desert had to be traversed without water, to the discomfort of men and beasts alike. The march was carried out under the heat of a burning sun. The suffocating dust kicked up by thousands of boots and hooves cocooned the column in its own local fog as it marched, and this was made worse by frequent sand-storms. By day the temperature soared to 110°F. By night it fell almost to freezing point. The record speed of the column, marching in such trying circumstances, is justly one of the Indian Army's most famous achievements.

As he marched, Roberts was joined by the detached garrison of Kalat-i-Ghilzai, 100 sabres of the 3rd Sind Horse, 2 guns of C–2 Battery, 2 companies of the 66th Foot, and the 29th Bombay Infantry (2nd Baluch Regiment). The whole force, preceded by their cavalry and played in by the bagpipes of the Highlanders, marched into the outskirts of Kandahar on 31 August.

On the approach of the relieving force, after heated exchanges in the Afghan camp between the leaders of the ghazis and the regular troops, Ayub abandoned the siege and fell back to his main camp about four miles north-west of the city.

Roberts now had at his disposal an army corps of two infantry divisions, two cavalry brigades, and six batteries of artillery (horse, field, heavy and mountain), in all about 14,000 regular troops. Another strong division of about 8,000 men under Phayre was expected to arrive within the week. Against this Ayub could pit only his one division of 5,000 regulars and not more than 10,000 ghazis and tribal irregulars, many without firearms, or armed only with jezails (Afghan muskets) and smooth-bore flintlocks.

On 1 September 1880 the British took up their positions for an attack. This time there were to be no mistakes. The troops were ordered to breakfast at 7 am, then store their kits behind walls to form strong-points in the rear. Divisional and brigade commanders reported for orders at 6 am. The Bombay troops from the Kandahar divisions were to hold the flanks and provide an escort for the 40-pounders of 5–11 Battery, now coming steadily out of the city drawn by their twenty yoke of oxen. The 1st and 2nd Brigades of the Bengal troops from Kabul were to make the main attack, sweeping round the edge of the range of hills which marked the front of the Afghan position, and outflanking them through the village of Pir Paimal on their right. The hill of Babawali Kotal, which marked the left of the Afghan position, was to be threatened by the 3rd Brigade. The centre was to be held by a strong detachment of troops from the Kandahar garrison, and the massed cavalry brigade formed up in the left rear, ready to operate in the open country behind Pir Paimal.

At about 9.30 am the 40-pounders opened fire on the Afghan battery occupying Babawali Kotal, which had been harassing the British camp the previous day. Now it was the Afghan gunners' turn to be outmatched in range and weight of metal. However, they made a spirited reply, and counter-battery bombardment continued while the rest of the attack proceeded on the other flank as the British pressed on towards Pir Paimal. The advance was made more difficult by the number of small villages, orchards, and wall enclosures through which the attackers had to pass, and severe hand-to-hand fighting took place before the Afghans holding them were dislodged and driven out. By midday both brigades, with artillery support, had fought their way round the Afghan flank, and were threatening to take their whole position in the rear.

The Afghans attempted to form a new line, at right-angles to their original position, with their left resting on Babawali Kotal. The guns there were swung round to meet the new threat, while the rest of the artillery formed up in the centre of the line, with their camp, ordnance park, and baggage behind them. The British pressed home their advantage. The defences in front of the camp were stormed by the 92nd Highlanders, 2nd Gurkhas and 23rd Pioneers. This brought them behind the original Afghan position and into the fire of 5–11, whose 'overs' fell too close for comfort. The chief signalling officer was killed here by a ghazi while trying to find an eminence from which to signal back the news of the British success. The battery on Babawali Kotal was rushed by the 3rd Sikhs, and the Afghan position there taken. Ayub Khan, seeing that all was lost, rode off to make his escape. His regular regiments dispersed in various directions, and the tribesmen made off to try to reach the safety of their own villages, there to resume the appearance of peaceful cultivators.

The infantry brigade halted to regroup and replenish their ammunition, in case the Afghans attempted to make another stand. However, no more serious resistance was encountered and by the time the cavalry came up the Afghans had made good their escape. A pursuit was made until nightfall, the 3rd Bombay Light Cavalry and 3rd Scinde Horse having their revenge for Maiwand by cutting down about 350 fugitive tribesmen before giving up the chase.

The entire Afghan camp was captured, with all thirty of Ayub's guns and the two from E–B Battery which had been lost at Maiwand. The body of Lieutenant Maclaine of the RHA, who had been captured in the retreat and had been treated as an 'honoured guest' by Ayub Khan, was also found. He had apparently been murdered by his guards or a ghazi after Ayub's flight. The other British losses in the battle came to 35 killed and 213 wounded. The Afghans were known to have lost more than 600 killed, from the bodies buried by the victors, and many more were killed in the retreat.

Ayub, his army scattered, disappeared from the political scene. Amir Abdul al-Rahman gradually extended his control over the country. The British agreed not to force an envoy upon him in return for a promise that Afghanistan would have no dealings with the Russians (who were as unpopular as the British and regarded with equal suspicion), and the last troops of the Indian Army were withdrawn in May 1881.

Maps

KEY TO THE BATTLE MAPS

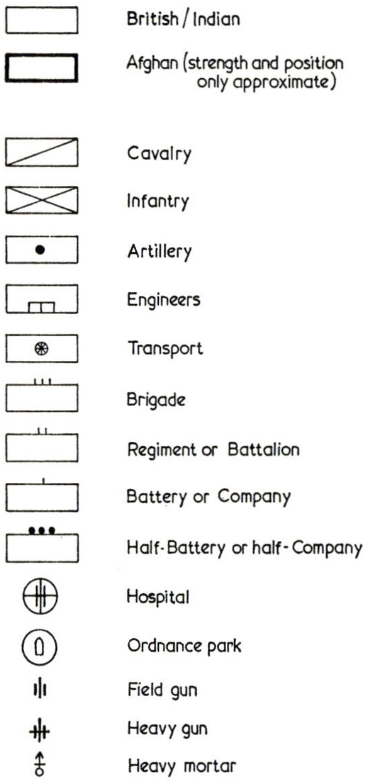

British / Indian

Afghan (strength and position only approximate)

Cavalry

Infantry

Artillery

Engineers

Transport

Brigade

Regiment or Battalion

Battery or Company

Half-Battery or half-Company

Hospital

Ordnance park

Field gun

Heavy gun

Heavy mortar

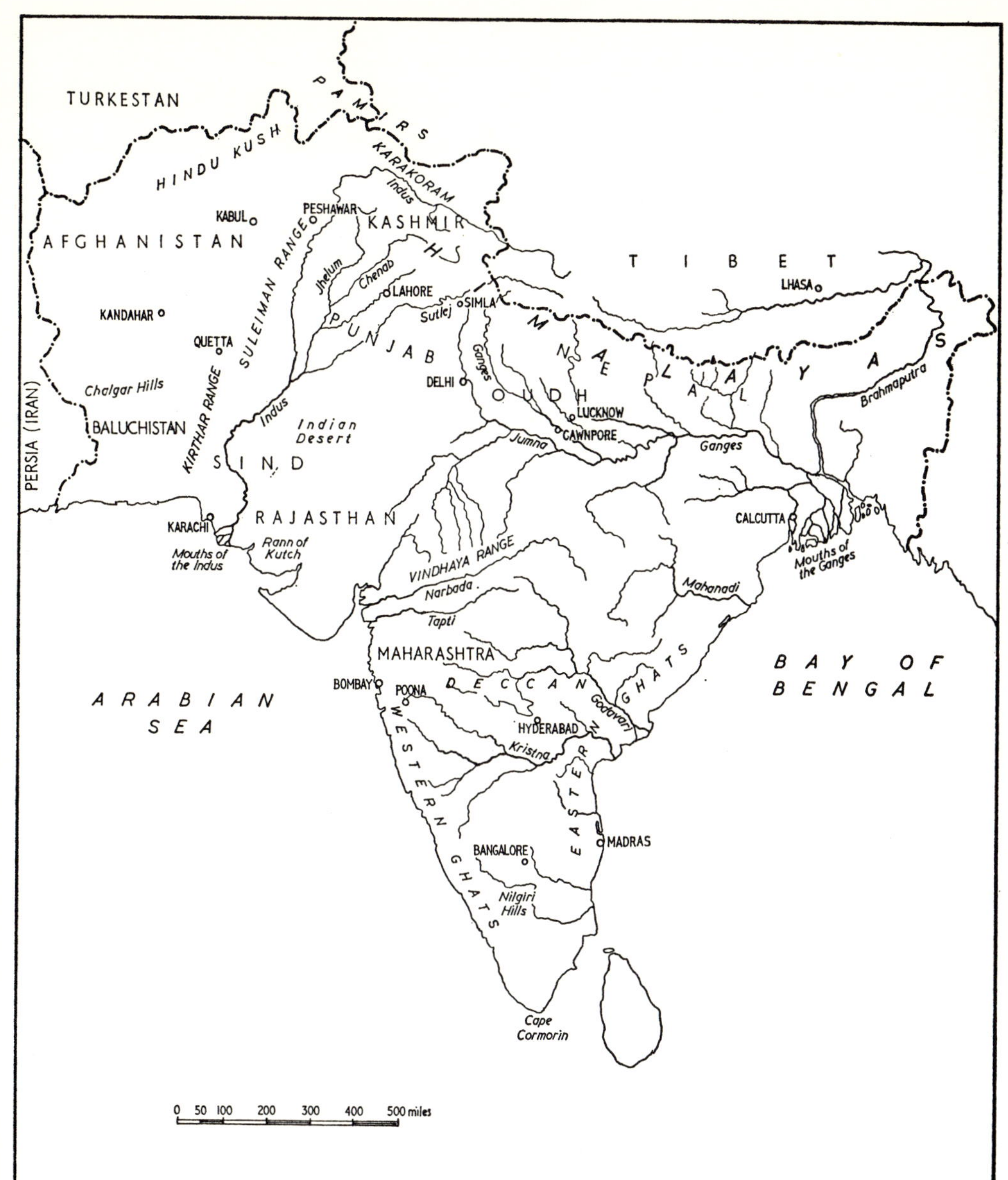

The Indian sub-continent: geographical features and important towns and provinces

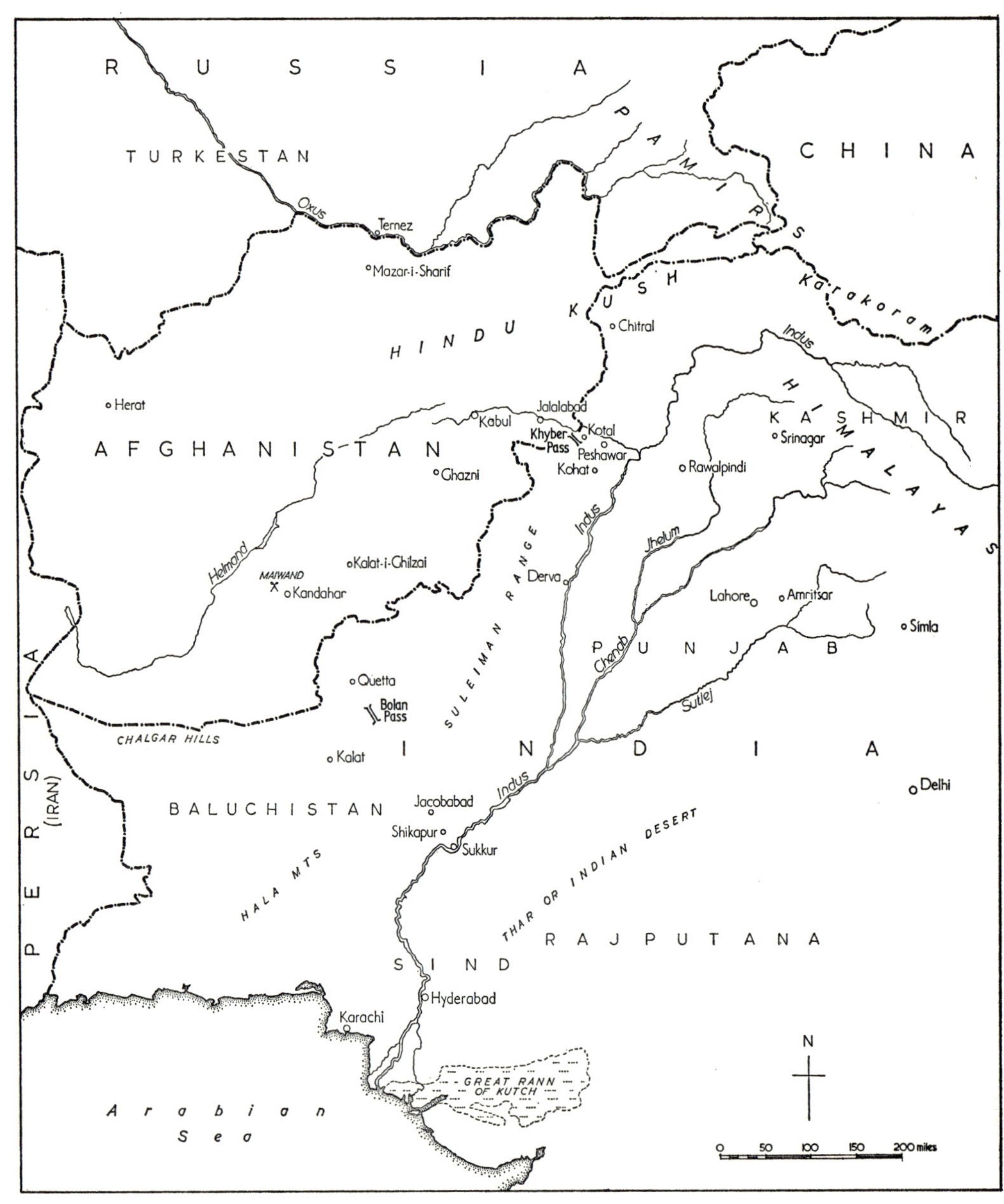

The North-West Frontier of British India c 1900

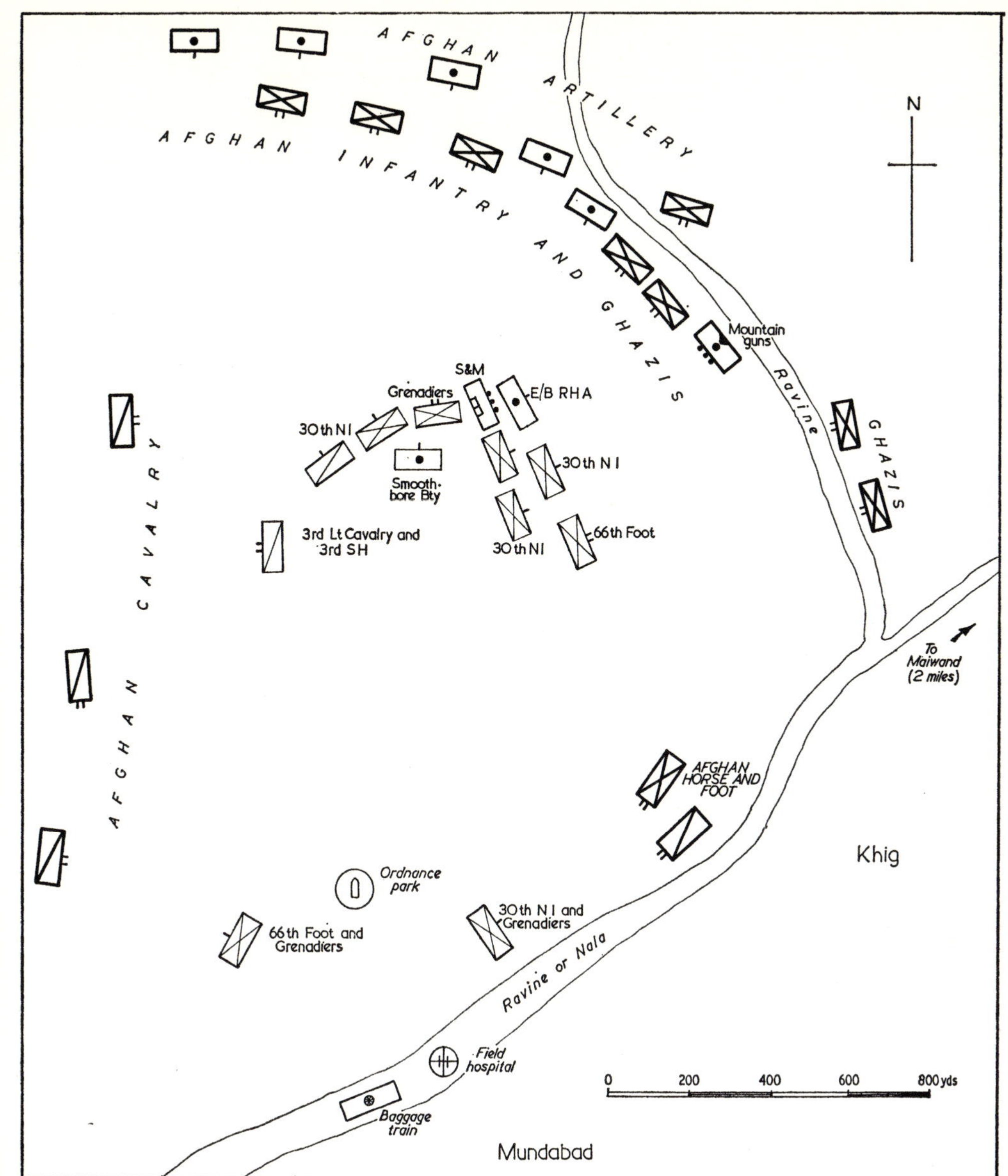

Battle of Maiwand, 27 July 1880: the position before the withdrawal of the Smoothbore Battery

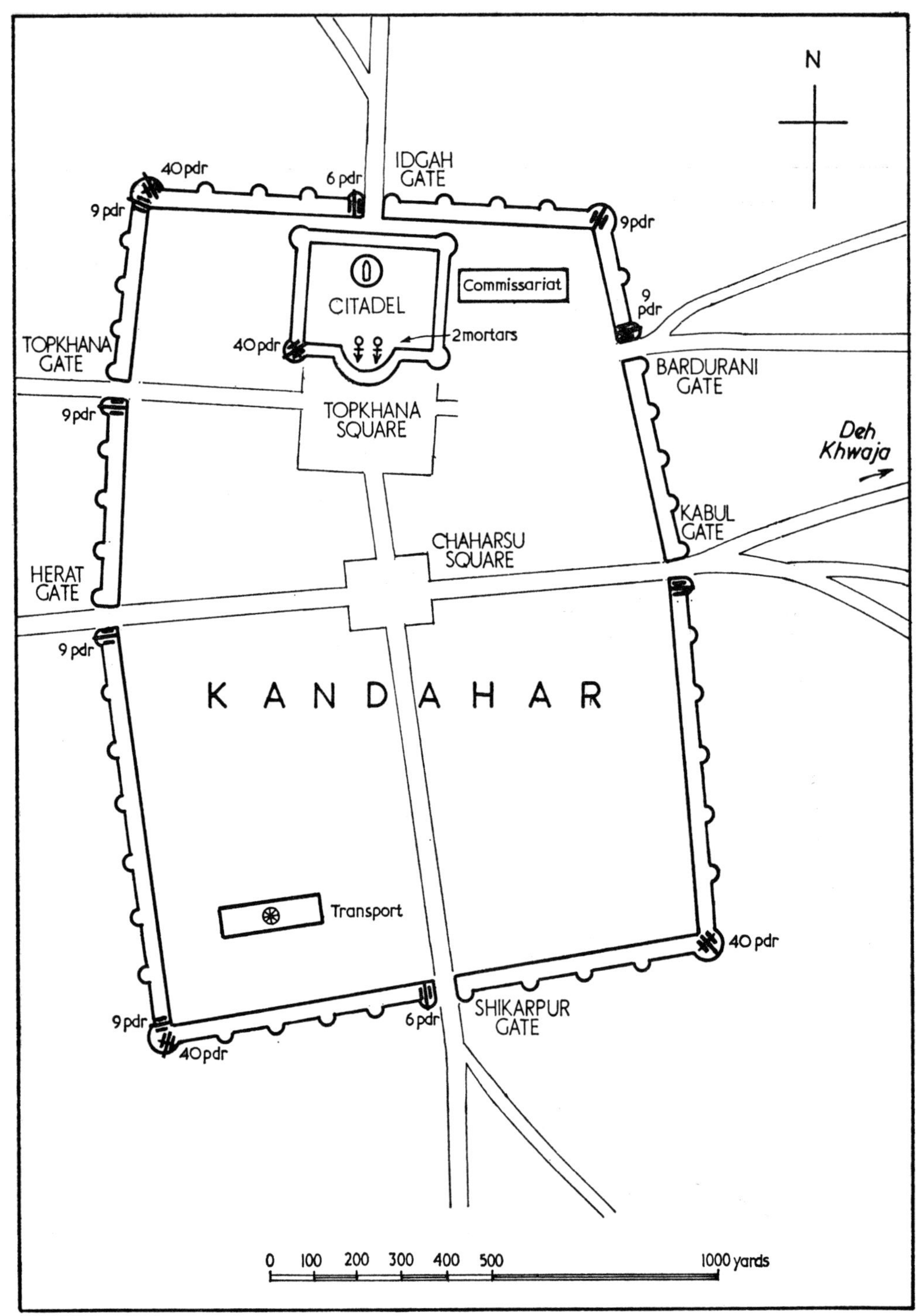

Kandahar, August 1880. Distribution of defending troops: 7th Fusiliers – South and East walls; 66th Foot – Idgah Gate and Citadel; 26th Bombay Native Infantry – North and West walls; 1st and 30th Bombay Native Infantry – Topkhana Square; 4th and 19th Bombay Native Infantry – Commissariat and Transport lines and the Chaharsu Square; 3rd Bombay Cavalry – Idgah Gate; Poona Horse – Topkhana Gate; 3rd Scinde Horse – north-west and south-west corners

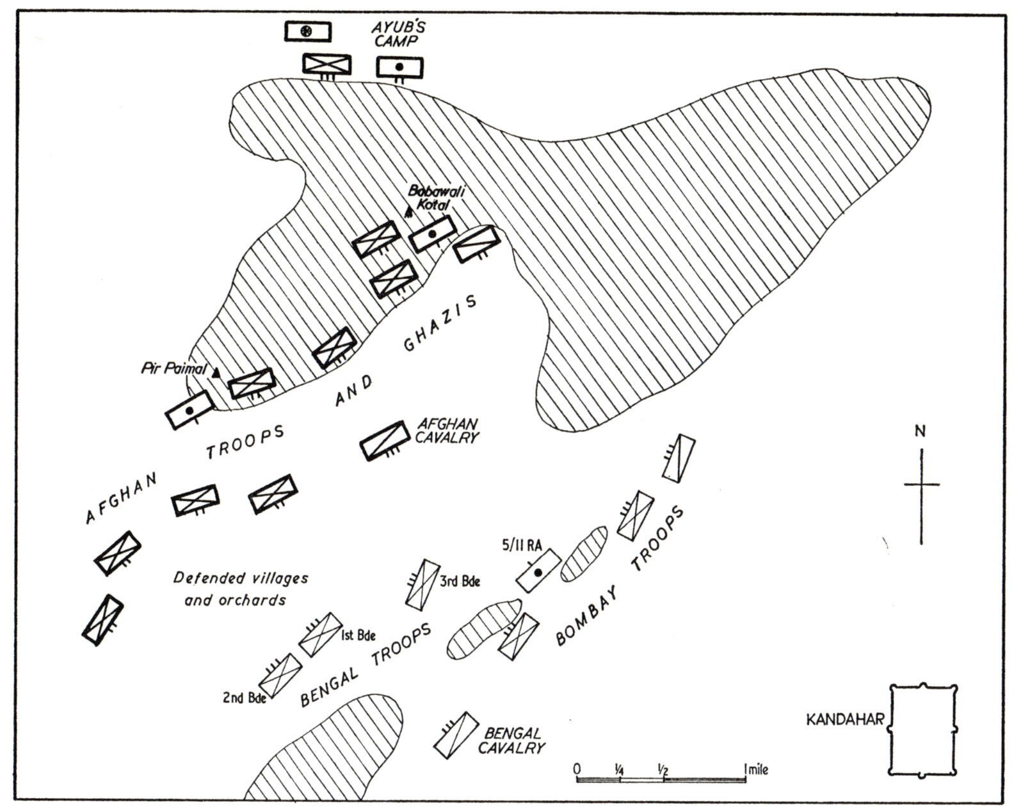

Battle of Kandahar, 1 September 1880: position at the beginning of the action. Shaded areas indicate high ground.

Appendices

Appendix 1

Local East India commission

His Excellency the Right Honourable HUGH, LORD GOUGH, Grand Cross of the Most Honorable Military Order of the Bath, etc. etc. General of Her Majesty's Forces, Commander in Chief of all the Queen's and Company's Forces in the East Indies, etc. etc. etc.

To Charles Raikes, Gentleman, 2nd Lieutenant in the Service of the East India Company.

By virtue of the Power and Authority in me vested by Her Majesty, and reposing especial Trust and Confidence in your Loyalty, Courage, and good Conduct, I do hereby constitute and appoint you the said Charles Raikes Gentleman, to hold the Rank of 2nd Lieutenant in the Queen's Army, in the East Indies only, and to take Rank as such from the 11th Day of June 1847. But as this Commission is granted to you in Virtue of the Rank which you bear in the Service of the Honorable East India Company, it is to have Force and Effect no longer than you shall remain in the said Company's Service, unless you shall be transferred with similar Rank into the immediate Service of Her Majesty. You are therefore carefully and diligently to Discharge the Duty of 2nd Lieutenant by doing and performing all and all manner of Things thereunto belonging; And I do hereby Command all Officers and Soldiers whom it may concern, to acknowledge and obey you as a 2nd Lieutenant in the Queen's Army, in the East Indies only, as aforesaid. And you are to observe and follow such Orders and Directions, from time to time, as you shall receive from Her Majesty, or any your Superior Officer, according to the Rules and Discipline of War, in pursuance of the Trust hereby reposed in you. Given under my Hand and Seal at Simla in Bengal, this Fifteenth Day of May, in the Year of our Lord One Thousand Eight Hundred and Forty Eight, and in the Eleventh Year of Her Majesty's Reign.

By His Excellency's Command,
(*signed*) Fred. P. Haines Capt
Military Secretary, East Indies

(*signed*) Gough General,
Commander in Chief,
East Indies

Appendix 2

Sepoys' enlistment oath

(Translation of the oath sworn on the regimental colours by sepoys on enlistment into the East India Company's army).

I, (full name) inhabitant of .... village, ... district, ... province, son of .... do swear that I will never forsake or abandon my colours; that I will march wherever I am directed, whether within or beyond the Company's territory; that I will obey all the orders of the officers set over me; and that I will in everything behave myself as becomes a good soldier and a faithful servant of the Company; and failing in any part of my duty, as such I will submit to the penalties described in the Articles of War which have been read to me.

Appendix 3

Order of battle of the Indian Army, 1857

British	*Units*	*Men*
Cavalry	4 regiments	2,686
Infantry (British Army)	22 battalions	21,577
Infantry (Company's army)	9 battalions	8,438
Horse Artillery	17 troops }	6,419
Foot Artillery	52 companies }	
Engineers	—	631
		Total British 39,751

Indian		
Regular Cavalry	21 regiments	9,876
Irregular Cavalry	33 regiments	21,047
Horse Artillery	6 troops }	4,166
Foot Artillery	36 companies }	
Regular Infantry	115 battalions	152,860
Irregular Infantry	45 battalions	35,426
Sappers and Miners	29 companies	3,043
		Total Indian 226,418

Appendix 4

Order of battle of the Indian Army, 1906

British	*Units*	*Men*
Cavalry	9 regiments	5,652
Horse Artillery	11 batteries	
Field Artillery	45 batteries	
Heavy Artillery	6 batteries	14,824
Mountain Artillery	8 batteries	
Garrison Artillery	22 companies	
Royal Engineers		210
Infantry	52 battalions	53,798
		Total British 74,484

Indian		
Cavalry	40 regiments	25,239
Mountain Artillery	11 batteries	
Frontier Garrison Artillery	1 Corps	7,099
Attached to Royal Artillery	—	
Sappers and Miners etc	33 companies	4,800
Commissariat Transport Corps		
Infantry and Pioneers	129 battalions	121,206
		Total Indian 158,344

Reserves		
British Volunteers	61 Corps	34,000
Indian Reservists		27,500
Imperial Service Troops	41 Corps	18,000
Military Police	21 battalions	17,500
Frontier Militia and levies		14,500
		Total Reserves 111,500

Glossary

Anna (*āna*) One sixteenth of a rupee (qv)
Asami (*asāmī*) A post; an appointment; the lump sum paid by a silladar (qv) to cover the cost of his horse and equipment
Crore (*Karōṛ*) (written out as 1,00,00,000) Ten millions (10,000,000) or 100 lakhs (qv)
Doolie (*Ḍōlī*) A litter; a covered stretcher
Dafadar (*Dafadār*) Indian sergeant of irregular cavalry
Gora-log (*Gōrā lōg*) White people; European soldiers
Havildar (*Ḥavāldar*) Indian sergeant
Havildar-Major Indian sergeant-major
Jemadar (*Jam'adār*) Indian assistant troop or company officer
Jawan (*Javān*) A young man, a man of military age, a soldier
Lakh, lac (*Lākh*) (written out as 1,00,000) A hundred thousand (100,000)
Loot (*Lūṭ*) Plunder; booty
Naik (*Nāyak*) Indian corporal
Paltan (*Palṭan*) Battalion (corruption from the French *peloton* = squad or platoon); hence, the Infantry
Ressala (*Risāla*) Troop or squadron; hence, the Cavalry
Ressaldar (*Risāldār*) Indian troop commander (irregular cavalry)
Ressaldar-Major The senior Indian officer of an irregular cavalry regiment
Rupee (*Rūpiya*) Indian silver coin, valued at one-tenth of a pound sterling (gold) until about 1870. The fall in silver against gold thereafter led to a fall in the value of the rupee, at times to one-twentieth of a pound. In the 1890s it was stabilised at fifteen to the pound, a rate which continued until the end of the British period.
Sahib (*Ṣāḥib*) Lord; master; a gentleman; Sir; Mr
Sepoy (*Sipāhī*) An Indian soldier; a private in an Indian regiment
Silladar (*Silāḥdār*) An Indian soldier providing his own horse and equipment
Sowar (*Savār*) An Indian cavalryman; a trooper; one who rides
Subedar (*Ṣūbadār*) An Indian company officer (also a troop officer of regular cavalry, until *c*1860)
Subedar-Major The senior Indian officer of a battalion
Top-Khana (*Tōp Khāna*) The cannon department; the artillery

Where a conventional English rendering of an Indian term exists, eg havildar, jemadar, doolie, I have used this in the text and given the more exact version in the Glossary. Similarly I have for the most part used the names of Indian cities and districts in the form in which they are most familiarly known to western readers (eg Punjab, not Panjāb, Oudh, not Āwadh). When in doubt I have used the form given in the *Imperial Gazetteer of India.* In the same way I have used a conventional spelling for personal names, and not troubled the non-specialist reader with intrusive diacretics – thus 'Khan' rather than 'Khāṇ'. Official military titles have been retained in their own spelling. Thus the *Scinde* Irregular Horse are described as defending the Upper *Sind* Frontier.

Note on Transliteration

The system of transliterating Urdu (Hindustani) words used throughout this book conforms to that of Sir William Hunter.

The vowels are rendered as follows:

a (short) – as u in butter
ā (long) – as in far
ay (long) – as in lay
e (short) – as in peg
i (short) – as in ink
ī (long) – as ee in sheep
o (short) – as in hot
ō (long) – as in bone
u (short) – as oo in book
ū (long) – as oo in boot

Most consonants are pronounced as in English, but without the 'aspiration' which they are usually given by English speakers.

Thus k is pronounced as the k in lock
but kh is pronounced as kh in blockhouse
Kh (guttural) is pronounced as ch in Scottish loch or German ich
Gh is, similarly, a guttural 'g' sound.

Retroflex consonants, ṭ, ḍ, etc, are pronounced by putting the tip of the tongue against the roof of the mouth. T and d are pronounced with the tip of the tongue between the front teeth.

Th is pronounced as in hothouse
c as the ch in church
ch as ch-h in church-hall

Diacretic marks are employed to indicate which of the various characters for h, s, a, etc, are used in the original Persi-Arabic script.

Bibliography

(A) DOCUMENTS

India Office Library and Records (Foreign and Commonwealth Office)
Private Papers of Sir John Lawrence (MSS Eur F90)
Private Papers of Edward, 1st Earl of Lytton (MSS Eur E218)
Private Indian Papers of Robert, 3rd Marquis of Salisbury (microfilm)
Private Papers of the Hon Mountstuart Elphinstone
Military Department Records: European muster rolls

Royal Military Academy Sandhurst
Cadet Registers (Royal Military College)
Private letters of Captain G. F. Paterson to his wife, 1914–19

School of Oriental and African Studies, University of London
Correspondence between John Jacob and Bartle Frere (No 138373)

National Army Museum
Private Papers of General Sir Henry Rawlinson
The Hodson Biographical Index of Indian Army Officers
The Narrative of Private Potiphar
The Journals of Private Hall

(B) PRINTED BOOKS

Aitchison, C. U. *Lord Lawrence* (1897)
Archbold, W. A. J. *Outlines of Indian Constitutional History* (British period) (1926)
Arthur, George. *Life of Lord Kitchener*, Vol 2 (1920)
Balfour, Lady Betty. *The History of Lord Lytton's Indian Administration 1876–1880* (1899)
Bandmaster, A. (pseud.). *Trumpet and Bugle Sounds for the Army* (undated)
Barnett, Corelli. *Britain and Her Army 1509–1970: A Military, Political and Social Survey* (1970)
Barrow, George. *The Life of General Sir C. C. Monro* (1931)
Basham, A. L. *The Wonder That Was India* (1954)
Betham, G., and Geary, H. V. R. *The Golden Galley: The Story of the Second Punjab Regiment, 1761–1947* (1956)
Bond, Brian. *The Victorian Army and the Staff College 1854–1914* (1972)
Buck, E. J. *Simla, Past and Present* (2nd ed, 1925)
Butler, W. F. *Life of Sir Charles Napier* (1894)
——. *Life of Sir George Pomeroy Colley* (1899)
Cadell, P. *History of the Bombay Army* (1938)
Candler, E. *The Sepoy* (1919)
Cardew, F. G. *The Second Afghan War, 1878–80* (abridged official account) (1908)
——. *A Sketch of the Services of the Bengal Native Army to the Year 1895* (1903)
Carman, W. Y. *Indian Army Uniforms under the British*, 2 vols (1961, 1969)
Carrington, C. *Rudyard Kipling: His Life and Work* (1955)
Carter, T. *Historical Record of the Forty Fourth, or the East Essex Regiment of Foot* (1864)
Casserly, Gordon. *Life in an Indian Outpost* (*c* 1910)
Churchill, Randolph S. *Winston S. Churchill*, Vol 1: *Youth (1874–1900)*; *Companion (1874–1896)* (1967)
Coen, T. C. *The Indian Political Service* (1971)
Cohen, S. P. *The Indian Army: Its Contribution to the Development of a Nation* (1972)
Colvin, Ian. *The Life of General Dyer* (1929)
Connell, John. *Auchinleck: A Biography of Field-Marshal Sir Claude Auchinleck* (1959)

Daniell, D. S. *4th Hussar (The Story of the 4th Queen's Own Hussars, 1685–1958)* (1959)
Delhi: Manager of Publications. *Our Indian Empire: A Short Review and Some Hints for the Use of Soldiers Proceeding to India* (1935)
Dilks, David. *Curzon in India*, 2 vols (1970)
Dodwell, H. H. (ed). *The Cambridge History of India*, Vols 5 and 6 (1929, 1932)
Dunbar, George. *Frontiers* (1932)
Dunsterville, L. C. *Stalky's Adventures* (1928)
East India Registers
Edwards, M. *A History of India* (1961)
Elsmie, G. R. *Field-Marshal Sir Donald Stewart* (1903)
Everett, H. J. *The History of the Somerset Light Infantry (Prince Albert's)* (1934)
Ferrar, M. L. *A History of the Services of the 19th Regiment* (the Green Howards) (1911)
Forrest, G. W. *Life of Field-Marshal Sir Neville Chamberlain* (1909)
Fortescue, J. W. *A History of the British Army*, Vols 12 and 13 (1927, 1930)
Godwin-Austen, A. R. *The Staff and the Staff College* (1927)
Graham, C. A. L. *The History of the Indian Mountain Artillery* (1957)
Gross, John. (ed). *Rudyard Kipling: The Man, His Work and His World* (1972)
Guest, Freddie. *Indian Cavalryman* (1959)
Guggisberg, F. G. *'The Shop': The Story of the Royal Military Academy* (1900)
Haidari, M. Akbar Khan. *'The Munshi': A Standard Hindustani Grammar* (6th ed, 1922)
Hanna, H. B. *The Second Afghan War, 1878–79–80: Its Causes, Its Conduct and Its Consequences*, 2 vols (1904)
Hannah, W. H. *Bobs: Kipling's General: Life of Field-Marshal Earl Roberts* (1972)
Hart's Army Lists
Headlam, J. *The History of the Royal Artillery 1860–1914*, 3 vols (1931, 1940)
Heathcote, T. A. *British Policy and Baluchistan 1854–1876*, University of London PhD History thesis (1970)
Hennell, R. *A Famous Indian Regiment: The Kali Panchwin* (5th Mahratta Light Infantry) (1927)
Henniker, M. C. A. *Memoirs of a Junior Officer* (1951)
Hensman, H. *The Afghan War of 1879–80* (1882)
Hodson, V. C. P. *List of the Officers of the Bengal Army 1758–1834*, 4 vols (1927–47)
Holmes, T. R. *Sir Charles Napier* (1923)
Hudson, H. *History of the 19th King George's Own Lancers, 1858–1921* (1937)
Ilbert, C. *The Government of India: Being a Digest of the Statute Law Relating Thereto* (1907)
Imperial Gazetteer of India, 25 vols (new ed, 1909)
India Office Army and Civil Lists
Jackson, D. *India's Army* (1940)
Jacob, J. *Record Book of the Scinde Irregular Horse* (1856)
James, David. *Lord Roberts* (1954)
James, H., and Sheil-Small, D. *The Gurkhas* (1965)
Jocelyn, J. R. J. *The History of the Royal & Indian Artillery in the Mutiny of 1857* (1915)
Joyce, Michael. *Ordeal at Lucknow: The Defence of the Residency* (1938)
Kerry, Stephen. *Doctor Sahib* (1967)
Kincaid, D. *British Social Life in India 1608–1937* (1938)
Kipling, John. *Beast and Man in India* (1892)
Lambrick, H. T. *John Jacob of Jacobabad* (1960)
——. *Sir Charles Napier and Sind* (1952)
Lawford, J. *30th Punjabis* (1972)
Lawrence, Rosamond. *Charles Napier: Friend and Fighter, 1782–1853* (1952)
Lee-Warner, William. *Memoirs of Field-Marshal Sir Henry Wylie Norman* (1908)
Lumsden, P. S., and Elsmie, G. R. *Lumsden of the Guides* (1899)
Lunt, James (ed and trans). *From Sepoy to Subedar: The Memoirs of Sita Ram Pandy* (1970)
Lunt, James. *Scarlet Lancer* (1964)
MacMunn, G. F. *The Armies of India* (1911)
——. *Behind the Scenes in Many Wars* (1930)

MacMunn, *Vignettes from Indian Wars* (1930)
——. *The History of the Sikh Pioneers* (1933)
——. *The Martial Races of India* (1935)
Magnus, Philip. *Kitchener: Portrait of an Imperialist* (1958)
Mainwaring, A. *Crown and Company (Historical Records of the 2nd Bn. Royal Dublin Fusiliers)*, Vol 1 (1911)
Masters, John. *Bugles and a Tiger* (1956)
Maunsel, E. B. *Prince of Wales' Own: The Scinde Horse* (1926)
Maurice, Frederick. *The Life of General Lord Rawlinson of Trent* (1928)
Merewether, J., and Smith, F. *The Indian Corps in France* (1919)
Mockler-Ferryman, A. F. *Annals of Sandhurst* (1900)
Montgomery, Field-Marshal Viscount. *Memoirs* (1958)
Murland, H. F. *Baillie-Ki-paltan (2nd Bn. Madras Pioneers) 1759–1930*
Napier, C. J. *Defects, Civil and Military, of the Government of India* (1857)
Napier, H. D. *Field-Marshal Lord Napier of Magdala* (1927)
Napier, W. *The Life and Opinions of General Sir Charles Napier*, Vols 3 and 4 (1857)
Norgate, J. N. (trans). *From Sepoy to Subedar: Memoirs of Sita Ram Pandy* (1873)
O'Malley. L. S. S. *The Indian Civil Service* (1931)
Pelly, L. *The Views and Opinions of Brigadier-General John Jacob* (1858)
Petre, F. L. *The Royal Berkshire Regiment (Princess Charlotte of Wales's)* (1925)
Philips, C. H. *The East India Company 1784–1834* (1940)
Pollock, J. C. *Way to Glory: The Life of Havelock of Lucknow* (1957)
Qureshi, M. I. *History of the First Punjab Regiment 1759–1956* (1958)
Rait, R. S. *Life of Field-Marshal Sir Frederick Paul Haines* (1911)
Rawlinson, H. G. *The History of the 3rd Bn. 7th Rajput Regiment* (1941)
Reynard, F. H. *The Ninth (Queen's Royal) Lancers 1715–1903* (1904)
Roberts, Field-Marshal Earl. *Forty-One Years in India: From Subaltern to Commander in Chief* (1897)
Royal Military College Record (magazine)
[Russell-Jones, P. (ed)]. *The Army in India: A Photographic Record 1850–1914* (1968)
Sands, E. W. C. *The Military Engineer in India*, 2 vols (1933–4)
Sen, Amiya. *The Structure and Organisation of the Bengal Native Infantry with Special Reference to Problems of Discipline,* University of London PhD History thesis (1961)
Sen, S. N. *Eighteen Fifty Seven* (1957)
Shadbolt, S. H. *The Afghan Campaign of 1878–1880*, 2 vols (1882)
Shand, A. I. *General John Jacob* (1900)
'Ship's Adjutant' (pseud.). *Ordered East: Advice on Procedure and Kit to an Officer on Transfer to India for the First Time* (1933)
Singh, Rajendra. *The Grenadiers* (1962)
Smith, H. B. *Life of Lord Lawrence*, 2 vols (1885)
Spear, P. *The Oxford History of India*, Vol 3 (1958)
Stephen, L., and Lea, L. (eds). *The Dictionary of National Biography*, 22 vols (1885–1937)
Stranks, C. J. (ed). *The Path of Glory: the Memoirs of John Shipp* (1969)
Thompson, E., and Garratt, C. T. *The Rise and Fulfilment of British Rule in India* (1934)
Tugwell, W. B. P. *History of the Bombay Pioneers* (1938)
Vibart, H. M. *Addiscombe: Its Heroes and Men of Note* (1894)
War Office Army Lists
Ward, S. G. P. *Faithful: The Story of the Durham Light Infantry* (1962)
Waters, R. S. *History of the 5th Bn. (Pathans) 14th Punjab Regiment* (1936)
Watson, W. A. *Central India Horse: The Story of a Local Corps* (1930)
Woodham-Smith, Cecil. *Florence Nightingale 1820–1910* (1950)
Woodyatt, N. *Under Ten Viceroys: The Reminiscences of a Gurkha* (1922)
Wylly, H. C. *Neill's 'Blue Caps'*, 3 vols (1923)
——. *Crown and Company*, Vol 2 (1923)
——. *The York and Lancaster Regiment 1758–1919* (1930)
Yeats-Brown, F. *Bengal Lancer* (1930)
——. *Martial India* (1945)

Acknowledgements

I should like to express my thanks to the members of the staff of the following institutions for their helpfulness and courtesy to me during the preparation of this work: the India Office Library and Records (Foreign and Commonwealth Office); the University of London Library; the Institute of Historical Research, University of London; the National Army Museum; the Royal Military Academy Sandhurst Library. I must also say how grateful I am for the help which I have been given by many former officers of the Indian Army, by British officers and soldiers who served in India, and by serving officers of the armies of India and Pakistan. I should also like to record a personal debt of gratitude to Mr H. P. Willmott, Senior Lecturer in the Department of War Studies at the Royal Military Academy Sandhurst, for reading the book in draft form and making many valuable suggestions for its improvement; Mr Robert Goodall, who photographed all the illustrations used in this book; Major Polly Walker, QARANC (retd) and Sister Louise Loughlin, of the Royal Military Academy Sandhurst Hospital, for advice on tropical medicine and hygiene; Mr Herbert Woodend, Superintendent of the Pattern Room, Royal Small Arms Factory, Enfield Lock, for advice on the types of muskets and rifles used; Miss Elizabeth Killerby, who did the arithmetic involved in working out statistical tables; Miss Marilyn Weeks, who typed my increasingly illegible manuscripts; and lastly my wife, who not only cheerfully accepted the conventional task of academic wives of reading the proofs and preparing the index, but also produced the statistics of the social origins of the Indian Army's officers.

Sandhurst T. A. Heathcote

Index

Names of regiments appear under the heading 'regiments', arranged in order of seniority within each army. Conditions of service etc of military personnel are listed under the headings 'officers', 'soldiers'. Page numbers in italic type indicate plates and maps.